Lehrstuhl für Elektrische Antriebssysteme
Technische Universität München

Study of AC-Side Soft Commutated PWM Three-Phase Converters

Ahmad Radan

Vollständiger Abdruck der von der Fakultät für Elekt[illegible] tionstechnik der Technischen Universität München zur Erlangung des akadem[illegible] schen Grades eines

Doktor–Ingenieurs

genehmigten Dissertation.

Vorsitzender: Univ.- Prof. Dr.–Ing. W. Boeck

Prüfer der Dissertation:

1. Univ.- Prof. Dr.–Ing. Dr.–Ing. h.c. D. Schröder
2. Univ.- Prof. Dr.–Ing. U. Wagner

Die Dissertation wurde am 18.11.1999 bei der Technischen Universität München eingereicht und durch die Fakultät für Elektrotechnik und Informationstechnik am 23.02.2000 angenommen.

Ahmad Radan

STUDY OF AC-SIDE SOFT COMMUTATED PWM THREE-PHASE CONVERTERS

ibidem-Verlag
Stuttgart

Die Deutsche Bibliothek - CIP-Einheitsaufnahme:

Ein Titeldatensatz für diese Publikation ist bei
Der Deutschen Bibliothek erhältlich

∞

Gedruckt auf alterungsbeständigem, säurefreien Papier
Printed on acid-free paper

ISBN: 3-89821-027-8

Printed in Germany

Acknowledgements

I wish to express my sincere gratitude to my Professor Dr.–Ing. Dr.–Ing. h.c. Dierk Schröder for the guidance, encouragement and caring support during my graduate study.

I am grateful to my Doctoral Committee: Professor Dr.–Ing. U. Wagner for his interest and expert opinion about this work and Professor Dr.–Ing. W. Boeck for presiding the committee.

I wish to thank Dr.–Ing. N. Fröhleke for reviewing this work and providing valuable suggestions for improving it.

I would like to thank staff of Institute of Electrical Drives of Technical University of Munich for providing enjoyable educational atmosphere.

I would also like to acknowledge the Iranian Ministry of Culture and Higher Education for providing the scholarship for me during studying in Germany.

Last, but not least, I have special thanks to my wife, Anahita Motamed Rezvani, for her patience, supports and understanding.

Munich, March 2000

Directions for reader

Topologies and modulation techniques are named according to their features in this work. After introducing a topology or a modulation technique in each section, the abbreviation of its name is often used in the following sections to avoid the repetition of long names. Appendix E provides a list of abbreviations used in this work for the convenience of the reader.

Additionally a glossary is provided in appendix D for the reader in order to clarify the phrases used in this work compared to the similar ones in other fields of power electronics.

Contents

1. Introduction

Three Phase PWM Converters have the capability of bi-directional power flow and controllable power factor with reduced current harmonics. Due to these capabilities, they are widely utilized in high power/performance AC-motor drives, utility interface and uninterrupted power supply applications as means for $DC \Leftrightarrow AC$ electric energy conversion in power range from a few watt to some megawatt in traction.

As shown in Fig. 1.1, the classic PWM converter has a relatively simple structure and generates a low frequency output AC voltage with controllable magnitude and frequency.

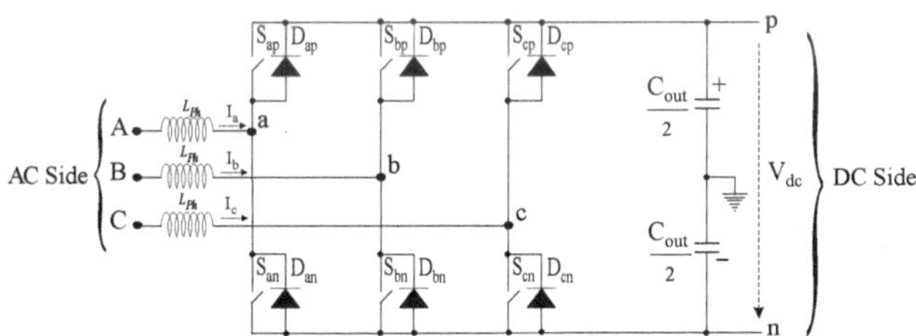

Fig. 1.1: Three Phase PWM Converter

As converter power switches, Mos Field Effect Transistors (MOSFETs) are used in some watt to some kilowatt applications. The medium power applications up to a few hundred KW can be mostly covered by Insulated Gate Bipolar Transistors (IGBTs). In the upper power range applications to some megawatt, Gate Turn-off Thyristors (GTOs) are well exploited.

The waveform quality of AC current is a point of great importance in $AC \Leftrightarrow DC$ converter systems. To increase the AC waveforms quality at reasonable costs, it is usually desirable to drive the converter at a switching frequency as high as possible. But, the maximum switching frequency of the classic PWM converter is strongly limited. The basic problem which limits the maximum switching frequency of the converter is the dissipated power in the semiconductors increasing with the increase of the switching frequency. With IGBTs the switching frequency can be increased to a few tens of kilohertz, while in the typical GTO applications this value is limited to a few hundreds of Hertz depending on the converter power rating. The hard switching PWM converters have also another problems related to high dV/dt and dI/dt when switching between the DC bus rails. High dV/dt produces EMI emission and causes failure in isolation materials of motor loads. High dI/dt results in the reverse-recovery problem of power diodes which furthermore increases the switching losses and the voltage stress on the switches. For these reasons improvement of the classic hard switching PWM converter in order to solve these problems has been the object of many research works for years and is the general object of this work too.

1.1 The Losses Related to Hard Switching PWM Converters

The losses propounded for a hard switching PWM converter can be divided into two terms: first, the losses of converter circuit and second, the losses caused by high order harmonics of the AC current in the loads or sources connected to the AC-side of converter. While the latter losses decrease with the increase of the switching frequency, the losses of converter circuit increase with its increase.

The converter circuit losses are dominantly determined by the device losses, which consist of conduction and switching losses. Conduction losses depend on the current through the device and the voltage drop across the device. Switching losses depend on the device current and voltage during the switching time. Generally speaking, with power MOSFETs the major switching losses occur during turn-on process and depend on the DC link voltage. This is because of the relatively large output capacitance and fast turn-off process of this device compared to bipolar devices. On the contrary, the major switching losses of IGBTs occur during turn-off process and depend on the turn-off current because of the tail part of the turn-off current which lasts a few microseconds even by today's fast IGBTs. The turn-on loss of this device is in return low due to its small output capacitance.

Switching losses increase proportionally with the increase of the switching frequency and become dominant at high switching frequencies. They limit the admissible maximum switching frequency of the PWM converters essentially. In order to reduce the switching losses, several technologies have been developed and published in research works. They will be briefly reviewed in the following sections.

1.1.1 Use of Snubber Circuits

Semiconductor power devices need snubber circuits because they have a limited safe operating area at turn-on and turn-off. The objective of a snubber circuit is to aid the switching devices during the switching transitions to withstand the voltage and the current stresses. These stresses are due to the interruption of the current at turn-off and to the collapse of the voltage at turn-on. The operating switching frequency bounds of any power devices can be extended by using the snubber circuits [1]-[3]. Electromagnetic interference (EMI) also decreases by using these circuits. Numerous topologies are introduced for snubber circuits [1]-[8]. The most effective snubber circuits are well-known as:

1. RCD-Snubber Circuit with overvoltage protection capacitor,
2. Asymmetrical Snubber Circuit (Undeland & Marquardt) and
3. Symmetrical Snubber Circuit (Wagner & McMurry).

They are shown for phase a of the converter in Fig. 1.2.

For the operation description and design consideration of the circuits see [12]. These circuits soften the switching conditions to some extent. However, despite the extra circuitry of the components, the voltage stress on the devices still remains at a level higher than the DC link voltage. Also the switching losses of the switches including the losses of their antiparallel diodes are transferred in part to their snubber circuits.

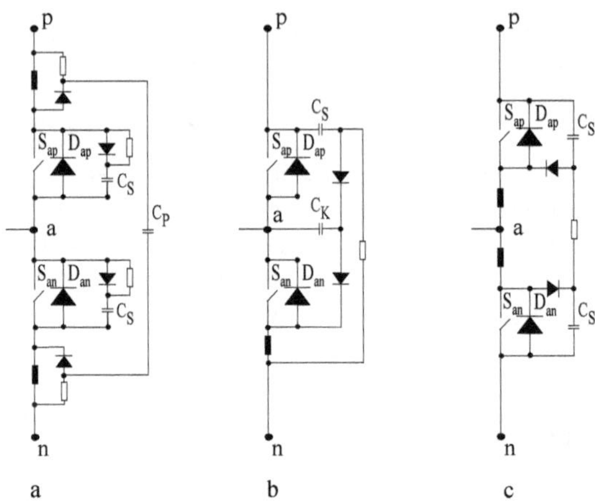

Fig. 1.2: a) RCD , b) Asymmetrical and c) Symmetrical snubber circuits for the switches S_{ap} and S_{an} lying at phase a.

From the design point of view the overvoltage protection capacitance C_P in the RCD-snubber circuit and C_K in the asymmetrical snubber circuit should be selected much larger than the capacitance C_S, in a scale of $30\mu F$ to $6\mu F$ for example, while the capacitance C_B in the symmetric snubber circuit can be selected as small as $3\mu F$ in the same example. The charging and discharging durations of these capacitances can result in a critical situation for small phase currents. Other dynamically critical situations can also appear due to the reverse and forward recovery effect of the snubber circuit diodes and the parasitic inductances of the circuits [12].

The inductors of the snubber circuits have to be selected large enough to limit dI/dt, so that the reverse recovery effect of the diodes and hence their losses are reasonably reduced. Since these inductors are located in the path of power flow and cause additional losses, this may not be desired in some applications.

1.1.2 Use of Soft Switching Techniques

For high power converters, IGBTs are more attractive than MOSFETs due to their lower conduction loss. The on-state voltage drop of IGBTs does not change so much with the anode current, so their conduction loss is almost proportional to their anode current. But Power MOSFETs have an ohmic on-state characteristic making their conduction loss proportional to square of their drain current. From the other hand, the slow switching characteristic of IGBTs, especially the turn-off tail current, causes severe switching losses limiting the increase of the switching frequency.

To exploit the advantage of low conduction loss of IGBTs, while keeping the switching frequency high enough, soft-switching techniques are employed to reduce the switching losses of IGBTs, especially their turn-off loss. For this purpose there exist two soft switching techniques, namely Zero Voltage Switching (ZVS) and Zero Current Switching (ZCS).

In the ZVS technique a capacitor is added in parallel to the IGBT to slow down the voltage rise during its turn-off transition. This reduces the voltage and current overlap and thus the turn-off losses. The added capacitor is charged during the turn-off transition and is then discharged during turn-on transition. In fact, a part of the turn-off losses is transferred to the turn-on instant by this capacitor. Without an auxiliary circuit the capacitor would be discharged through the switch resulting more turn-on losses. For solving this problem, an auxiliary circuit discharges the added capacitor by reducing the voltage across the switch to zero before the switch is turned on. This drastically reduces the turn-on losses. This technique is more effective when it is applied to a fast switching IGBT with a relatively small tail current, for the turn-off losses are reduced only in part. On the contrary, the ZCS technique eliminates the voltage and current overlap during the turn-off transition by forcing the switch current to zero before the switch is turned off. This technique does not reduce the turn-on losses, it may even result in an increase in the turn-on losses as it will be shown in an example in chapter 2. Thus ZCS is deemed more effective in reducing switching turn-off loss for slow devices where the major losses occur during turn-off. Since ZVS technique, as it will be shown in chapter 2, not only solves the problems related to reverse recovery effect of power diode, but also eliminates the turn-on loss and reduces the turn-off loss at the same time, it seems to be more attractive for today's fast IGBTs. That is why this work targets the circuits employing the ZVS technique.

1.2 Soft Commutated Converters

In most zero voltage switching technologies, or as named in this work "Soft Commutating Technologies", an auxiliary resonant circuit provides the soft switching conditions. In recent years, a number of soft commutated converters have been proposed. Unfortunately, switching losses in many of these circuits can be reduced only at the expense of much increased voltage/current stress on the devices, which lead to substantial overrating of the converter. In case of active auxiliary circuits, the switches of the auxiliary circuit operate mostly under increased voltage stress and hard switching conditions causing additional losses in the auxiliary circuit. Some of these circuits need a special PWM-technique for auxiliary circuit operation, which is not optimal in view of AC current quality or full utilization of DC bus voltage.

Soft commutated converters can be basically classified into two groups: DC-Side Soft Commutated Converters and AC-Side Soft Commutated (ASSC) PWM Converters. With the phrase "DC-side" it is meant that the soft commutating conditions are provided by reducing the DC link voltage, V_{dc}, to zero at the DC-side nodes p and n. In the AC-side concept, on the contrary, these conditions are provided by reducing the voltage across the on-going switches to zero at the AC-side nodes a, b and c.

The DC-side soft commutated converters are known as resonant converters. All conventional resonant converters suffer from significantly increased current or voltage stresses compared to PWM converters. They require excessive power ratings of resonant components and semiconductor devices, which reduce the converter efficiency and make them uneconomical [13]- [18].

To overcome the most serious problems of the conventional resonant converters, the resonant DC link (RDCL) converter was developed by shifting the resonant branch to the DC-side [20] (Fig. 1.3(a)). This converter provides zero voltage switching conditions for all main switches at the expense of two-fold increase of device voltage stress as compared to PWM converters. Additionally,

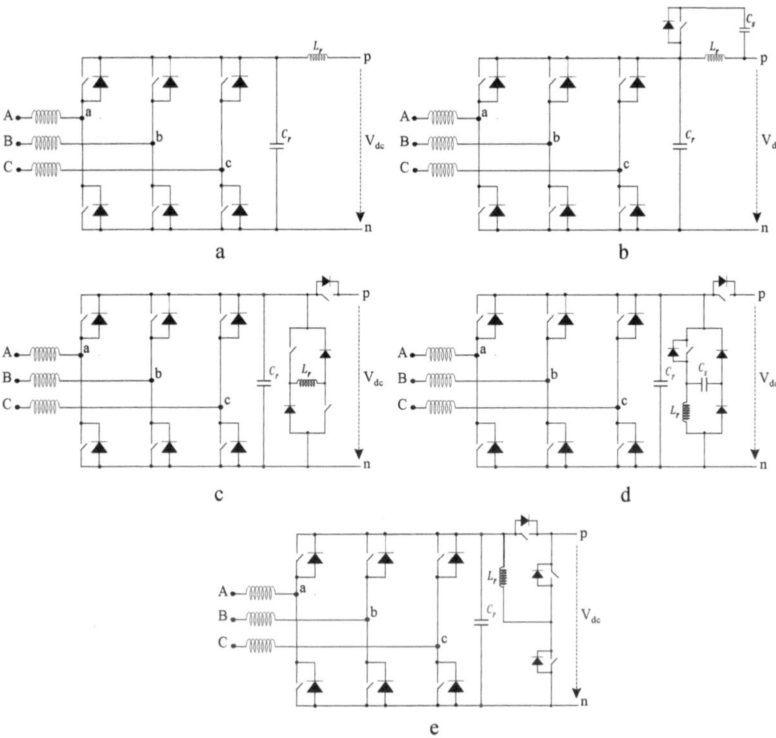

Fig. 1.3: Resonant DC link converters a) conventional RDCL converter, b) actively clamped RDCL converter, c, d and e) parallel (quasi) RDCL converters.

all the converter switches are required to switch synchronously with the resonant DC link. This requirement significantly reduces the converter's ability to achieve fine resolution for synthesizing the AC voltage and current waveforms at high power levels, for the attainable DC link resonant frequency is often not sufficiently higher than the frequency of the synthesized waveform. Many enhancements of the RDCL converter were introduced to alleviate these problems. Among them, the actively clamped resonant DC Link (ACRDCL) converter is considered to be the most successful topology [21] (Fig. 1.3(b)). The ACRDCL technique reduces the device voltage stress down to 1.2-1.4 times that of the

PWM converters. However, it is still higher than the DC link voltage, V_{dc}. It has also several disadvantages that come from discrete pulse modulation (DPM), which causes a subharmonic problem. The conduction loss in resonant operation is another problem of this circuit [30]. The enhancements of RDCL converter also resulted in the development of control strategies capable of providing waveforms with a quality comparable to those of PWM converters. (see also [22]-[25]).

Recognising that the resonant components should not be involved in the major power transfer processes and should only function during the switching transients, the resonant components were shifted to a path in parallel with the DC link, so the parallel resonant dc link (PRDCL) technique, also named the quasi-resonant DC link (QRDCL) technique, was proposed [26]-[30] (Fig. 1.3(c, d and e)). The circuits utilizing this technique reduce the device voltage stress to the lowest possible value, which is the DC link voltage V_{dc}. However, they introduce auxiliary switches which process up to 50% of the total output power of converter. This significantly reduces the converter efficiency and reliability. Their PWM capability is also limited because the converter switches should be turned on synchronously. Generally, this fact that some auxiliary devices have to be placed in the main power path is a common disadvantage of DC-side commutated converters. This can be overcome by placing the soft commutating circuit at the AC-side. For this reason, the AC-side soft commutating is superior to its DC-side concept in medium and high-power applications.

1.3 Goal and Contents of This Work

Much research has been done on the DC-side soft commutated converters but up to now the AC-side soft commutated converters have not been studied and analysed in detail. Also they have not been compared with each other in any work. The requirements of the auxiliary circuit of ASSC PWM converters limiting the degrees of freedom of modulation have not been determined precisely. The effect of duration of switching transitions on the DC bus utilization and maximum linear modulation index have not been discussed in the past.

The objectives of this work are to reexamine the properties of the most effective and recently proposed ASSC PWM converters, to analyse them precisely, to compare them with each other and also to improve their topology and modulation technique in order to minimize the losses of the whole circuit while keeping the quality of AC currents on an acceptable level.

Chapter 2 of this work starts with a discussion about the different soft switching techniques. Three single phase circuits operating in hard switching, ZVS soft switching and ZCS soft switching conditions will be considered. The behaviour of IGBT under the different switching conditions will be discussed based on the results obtained from the simulations using accurate physics-based dynamic models of IGBT and power diode. The simulation results will also be used for calculation of the device losses under the different switching conditions. This chapter will be continued with reviewing the ASSC PWM converters. Four concepts are proposed in this work. These concepts are the most attainable topologies in three phase high power applications. The operation principles of the proposed converters will be described based on the results obtained from the simulation of their circuit using simple models of IGBTs and power diodes without considering dynamic behaviour of devices. The modifications that should be necessarily carried out for conformity of the modulation technique with the requirements of auxiliary circuit operation will also be discussed for each concept in this chapter. The well-known auxiliary resonant commutated pole circuit [31] will be slightly modified for simplifying the timing control of main and auxiliary switches. The modified circuit is introduced as AC-Side Half-Wave Auxiliary Resonant Commutated Pole (ACS-HW-ARCP) Converter.

The ASSC PWM converters will then be analysed in detail in chapter 3. Beside the design equations, the expressions describing device voltage and current stresses will be derived for the converters under study. A comparison regarding the switching and auxiliary circuit energy losses of converters will also be given in this chapter.

In order to approach the main goal of this work, namely minimization of the entire circuit losses, different modulation techniques will be investigated in chapter 4. A detailed description of continuous and discontinuous space vector modulation techniques will be given. For the pole commutated circuits, the only circuits that can be controlled by any switching pattern, the modulation technique will be improved to minimize the number of switching actions and auxiliary circuit interventions. The improved technique will be introduced as Minimum-Loss Space Vector Modulation. The global and local RMS values of current ripples, as a measure of waveform quality for the AC currents, will be calculated for the proposed technique and will be compared with that of other techniques under the same switching losses.

The key issues of limitations of duration of voltage vectors in continuous and discontinuous space vector modulations resulting in reduction of the linearity range of modulation and of the available area of the voltage hexagon of space vector modulation will be discussed in chapter 5. This chapter will also deal with the effect of duration of switching transitions on the reduction of DC bus utilization and will suggest a method for compensating these effects.

In chapter 6 a dead-beat current regulator, which is suitable for the ASSC PWM converters operating with the minimum-loss space vector modulation, will be explained. The detailed current control block diagram of a converter system including an AC/DC converter, a DC/AC converter and an AC electric machine will be given for implementation in a DSP based controller.

2. AC-Side Soft Commutated PWM Converters

In many applications, it is desirable to have a controlled converter that can provide the following features: regulated and undistorted voltages and currents at both the AC and DC side, minimal number of components, high efficiency, high power density, and low electro-magnetic interference (EMI). For high power application, a 3-phase, PWM converter, shown in Fig. 2.1, is the most frequently used topology.

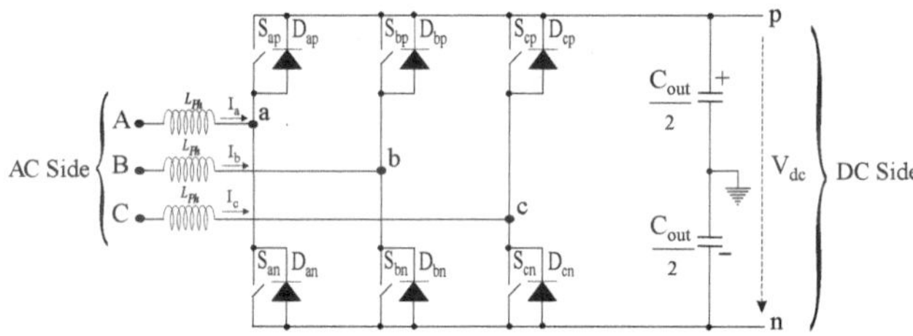

Fig. 2.1: Three-Phase PWM Converter.

As described in chapter 1, this circuit suffers from several problems that reduce circuit efficiency, produce high EMI emission and limit the switching frequency. To overcome these problems some kind of soft commutating techniques are developed and implemented. This chapter reviews the AC-side soft commutating techniques along with their implementing circuits. Since the Space Vector Modulation (SVM) technique is used today for switching control of almost all three phase PWM converters and often should be modified to meet the requirements of the soft commutating circuits, a brief review about the standard SVM will be given first. The different SVM techniques will be discussed later in chapter 4.

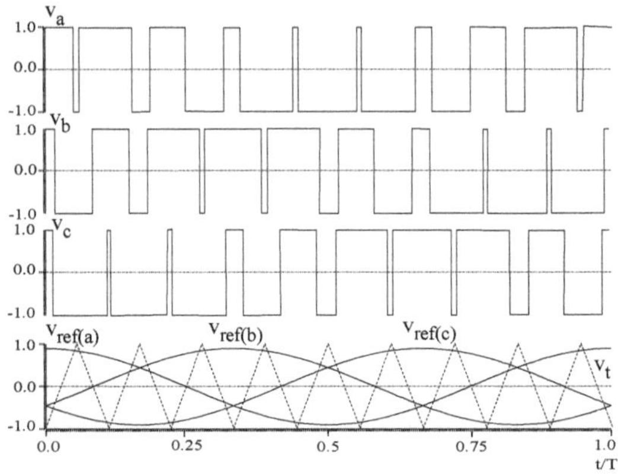

Fig. 2.2: Normalized waveforms of sinusoidal PWM.

2.1 Modulation Technique

The most common modulation scheme for three phase converters is Pulse Width Modulation (PWM). Normally, this would be a simple sinusoidal PWM technique, in which the reference sinusoidal waves, $V_{ref(a)}$, $V_{ref(b)}$ and $V_{ref(c)}$, are compared with a carrier triangular wave, V_t, and their crossing points determine the turn-on and turn-off time of the switches as shown in Fig. 2.2. In this figure the voltage axes are normalized to $V_B = V_{dc}/2$ and the time axis is normalized to the period of reference voltages T_1.

In contrast to the sinusoidal PWM method, the space vector modulation technique treats the three phase converter as a single entity and is based on the transformation of three-phase quantities into the (α/β) plane. The switches of the power converter are driven to obtain the desired converter switching states. The switching states consist of six non-zero states, $S_{1(pnn)}$ to $S_{6(pnp)}$, and two zero states, $S_{0(nnn)}$ and $S_{7(pnp)}$, as shown in Fig. 2.3(a). The switching state $S_{1(pnn)}$, for example, means that nodes a, b and c are connected to the positive, negative and negative DC rail voltage respectively. These connections can be either through a closed switch or through a conducting diode.

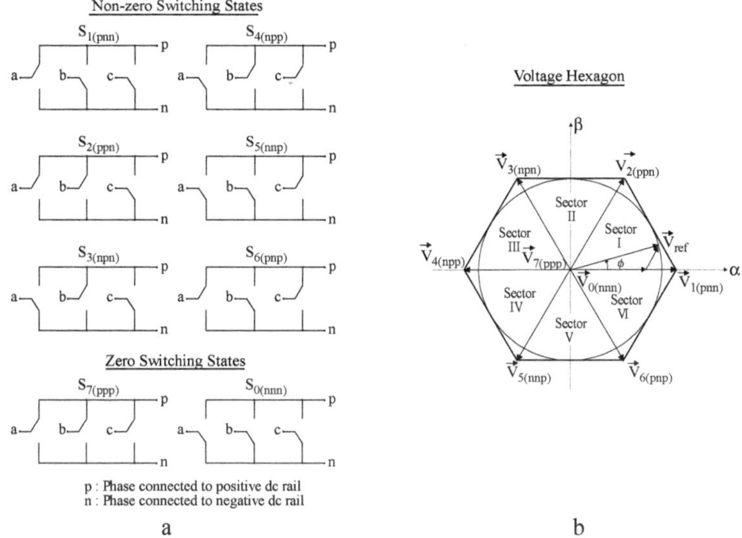

Fig. 2.3: a) Converter switching states and b) voltage vectors produced by the switching states and the voltage hexagon in SVM.

The converter switching states produce six non-zero voltage vectors, $\vec{V}_{1(pnn)}$ to $\vec{V}_{6(pnp)}$, which can be plotted on the (α/β) plane as six active vectors, and two zero voltage vectors, $\vec{V}_{0(nnn)}$ and $\vec{V}_{7(pnp)}$. The active vectors form a voltage hexagon and two zero vectors are located at the origin (Fig. 2.3(b)). The desired three phase sinusoidal reference voltages $V_{ref(a)}$, $V_{ref(b)}$ and $V_{ref(c)}$ with the amplitude $|V_{ref}|$ and the frequency $\omega_1 = 2\pi/T_1$ produce a rotating reference voltage vector $\vec{V}_{ref}$ in the (α/β) plane with the length $|V_{ref}|$ and the angle $\phi = \omega_1 t$. The locus curve of the trajectory of this vector corresponds to a circle in the (α/β) plane as shown in Fig. 2.3(b). The reference voltage vector $\vec{V}_{ref}$ is sampled in discrete time steps, which is called sampling cycle T_S[1)]. In each sampling cycle T_S, the required reference voltage vector $\vec{V}_{ref}$ is generated by switching its two directly adjacent active voltage vectors whose duty cycles are determined in such

[1)] We distinguish between the sampling and switching cycles. The switching cycle T_{sw} within which each one of the modulated switches is once turned on and off, is not equal to sampling cycle for most modulation techniques. (see chapter 4)

a way that the average of the real output voltage vector over one sampling cycle coincides with the reference vector. When the reference vector $\vec{V}_{ref}$ is located in sector I, as shown in Fig. 2.3(b) for example, it is generated by voltage vectors $\vec{V}_{1(pnn)}$ and $\vec{V}_{2(ppn)}$. Their duty cycles T_1[2] and T_2 are given by [61]:

$$\begin{aligned} T_1 &= T_S \frac{\sqrt{3}}{2} m \sin\left(\frac{\pi}{3} - \phi\right), \\ T_2 &= T_S \frac{\sqrt{3}}{2} m \sin(\phi), \end{aligned} \tag{2.1}$$

where

$$m = \frac{|\vec{V}_{ref}|}{\frac{V_{dc}}{2}}, \tag{2.2}$$

is the modulation index.

The sum of the two duty cycles T_1 and T_2 is smaller than T_S as long as the reference vector $\vec{V}_{ref}$ is inside the voltage hexagon. The remaining duration has to be completed by the zero vectors.

$$T_{0/7} = T_S - T_1 - T_2. \tag{2.3}$$

In the example of Fig. 2.3(b), $\vec{V}_{ref}$ is synthesized as follows:

$$\vec{V}_{ref} = t_1 \vec{V}_1 + t_2 \vec{V}_2 + t_{0/7} \vec{V}_{0/7}, \tag{2.4}$$

where $t_1 = T_1/T_S$, $t_2 = T_2/T_S$ and $t_{0/7} = T_{0/7}/T_S$ are the normalized duty cycles.

The maximum sinusoidal output voltage corresponds to the largest circle surrounded by the voltage hexagon (see Fig. 2.3(b)), so the fundamental crest value of the maximum line-to-neutral voltage is given by $V_{dc}/\sqrt{3}$. This results a maximum modulation index equal to $m_{max} = 2/\sqrt{3} = 1.15$, which is 15% more than that of the sinusoidal PWM.

The switching sequence of the voltage vectors and selection of the zero vector are the freedom degrees of the space vector modulation. They do not affect the

[2] It should not be confused with the fundamental cycle T_1.

resulting line-to-line AC voltage, but strongly influence the switching losses and AC current ripples, a subject to be discussed in chapter 4 where the different SVM techniques will be introduced.

2.2 Switching Techniques

In this section, the well known switching techniques: Hard Switching (HS), Zero Current Switching (ZCS) and Zero Voltage Switching (ZVS) will be reviewed. Three circuit diagrams are selected as example to introduce these techniques and to show how the problems of hard switching can be solved by soft switching techniques. The switching losses of IGBTs, as a major used power switch, under these switching conditions will be discussed. A comparison in view of total switching losses including the reverse-recovery losses of diodes will also be given. In the discussion of this section the gate circuit losses will be ignored.

2.2.1 Hard Switching

The circuit of figure 2.4(a) can represent phase a of the three phase converter of Fig. 2.1 during one switching cycle T_{sw}. In this single phase model it is assumed that phase current I_a flowing through diode D_{ap} is commutated to the IGBT power switch S_{an} at t_1 and then is commutated back from S_{an} to D_{ap} at t_3 under hard switching conditions. The operating waveforms of the circuit during one switching cycle are shown in Fig. 2.4(b). The waveforms are expanded for the turn-on and turn-off intervals in Fig. 2.4(c). The phase current is assumed to remain constant within the switching cycle. The inductors L_1 and L_2 represent the parasitic inductances of the connecting wires between the converter and the DC bus rails.

Suppose at t_1, while diode D_{ap} is conducting the phase current[3)] , S_{an} is turned on. During the turn-on transition $t_1 - t_2$ there is a large overlap between the switch current and voltage, $I_{S_{an}}$ and $V_{S_{an}}$. This results in the high turn-on loss of the switch as shown in the waveform of switch energy loss $E_{S_{an}}$ in Fig. 2.4(c).

3) When diode D_{ap} is conducting, its antiparallel switch S_{ap} can be closed or open. It depends on the application. Closing the switch ensures the phase voltage to be clamped to the DC rail, even if the phase current changes its direction while it is crossing the zero. Leaving it open avoids a number of switchings, which are of course zero current switching. If S_{ap} is closed during the conduction of D_{ap}, it should be turned off before the turn-on time of S_{an}.

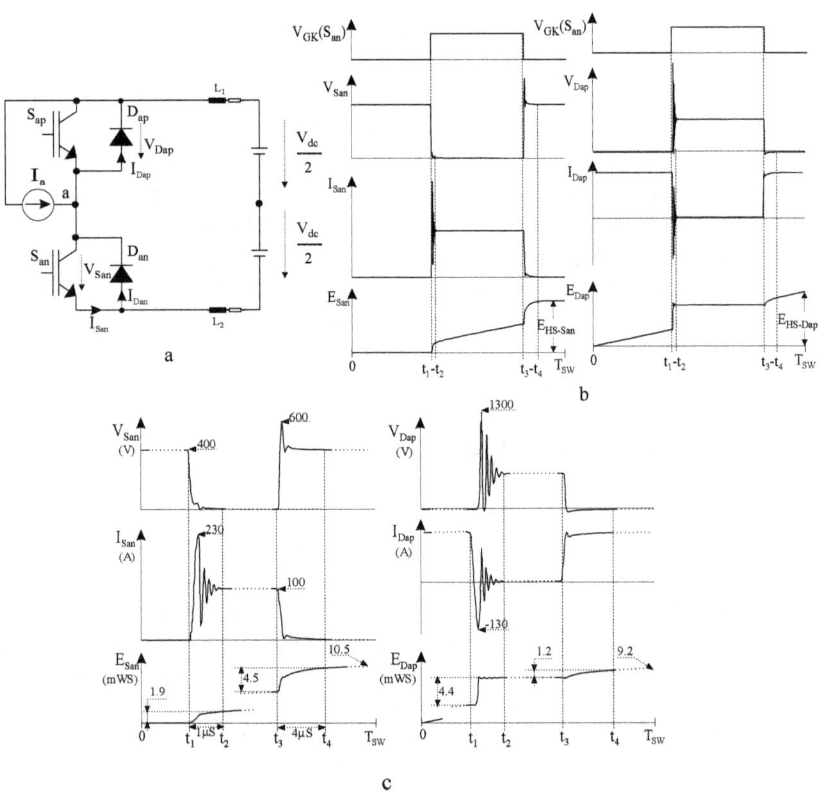

Fig. 2.4: Hard Switching (HS): a) circuit diagram b) operating waveforms c) expanded waveforms.

The turn-on loss consists of the loss associated with the establishment of the conductivity modulation[4)] and the loss caused by the reverse-recovery of the upper diode D_{ap}. The switch current rise time becomes longer because of the diode reverse-recovery, for much higher carrier modulation level has to be established. This leads to a considerable increase in the turn-on loss. As shown in the waveform of diode energy loss $E_{D_{ap}}$ in Fig. 2.4(c), the loss caused by the reverse recovery effect in the diode is also considerably high.

The parasitic inductances of the circuit are another problem. The resulting resonance among the parasitic inductances, L_1 and L_2, and the junction capacitance of diode D_{ap} during $t_1 - t_2$ as shown in Fig. 2.4(c) leads to undesired oscillations that increase the thermal stress of the diodes.

During the turn-off transition, $t_3 - t_4$, the current in the switch S_{an}, $I_{S_{an}}$, is cut off. Phase current I_a charges the parasitic capacitances across the switch S_{an} and discharges those across the switch S_{ap}. When the voltage of node a reaches the positive DC rail voltage, diode D_{ap} starts conducting. After the initial transfer of the switch current, $I_{S_{an}}$, to the upper diode D_{ap}, the remaining tail current decreases slowly. Its decrease is determined predominantly by the recombination of existing minority carriers. The tail current lasts for several micro-seconds even for very fast IGBTs available today and is the main cause for the turn-off loss. During the turn-off time, $t_3 - t_4$, the forward recovery effect of the diode D_{ap} and the parasitic inductances of the circuit cause a high voltage spike on the switch which can damage the switch.

The problems related to high di/dt at turn-on can be solved in high power IGBT converters by using the gate-controlled di/dt limitation technique [32]. This technique uses the internal stray inductance of the power module to sense the current and has no need for any extra component on the AC-side. But, even small inductances located on the main power path cause the problem related to parasitic inductances described above. The switch turn-on energy loss is also increased by using this technique. The zero voltage switching technique, which will be described in the next section, needs some extra components on the AC-side (not on the main power path), but solves the problems of turn-on transient more effectively.

[4)] Carrier injection into the drift region.

High voltage spike and high dv/dt on the switch at turn-off can be avoided either by using gate-controlled dv/dt limitation technique [32] or by adding the snubber capacitors. The former needs a small external high voltage capacitor to sense the voltage transients and increases the turn-off energy loss except in very high power applications. The latter reduces the turn-off energy loss but increases the turn-on energy loss. So it is also preferable to use the snubber capacitors with the zero voltage switching technique reducing the turn-on loss.

2.2.2 Zero Voltage Switching

The zero voltage switching technique was first introduced to solve the capacitive turn-on problem of power devices and the reverse recovery problem of the slow diodes [34]. It is considered most desirable for high-frequency power conversion with power devices whose main source of switching losses is the capacitive turn-on loss.

The circuit diagram of Fig. 2.4 with a ZVS scheme, along with its operation waveforms, are shown in Fig. 2.5. In this circuit an extra resonant branch (S_x, L_x and C_s) is added to create the ZVS-conditions for the switch S_{an}. The auxiliary branch is only activated prior to the turn-on of this switch and commutates the current I_a from D_{ap} to S_{an}. Suppose again the same initial conditions as in case of hard switching, i.e. diode D_{ap} is conducting the phase current, I_a. In this circuit the switch S_{ap} should be on when D_{ap} turns off. This is because of the operation of the resonant branch described later in this section. At t_1, prior to the turn-on of the power switch, S_{an}, the auxiliary switch, S_x, is turned on. The voltage $V_{dc}/2$ is applied across the resonant inductor, L_x, causing its current to increase linearly. The resonant branch reduces the current of the upper diode, $I_{D_{ap}}$, during the time $t_1 - t_2$ slowly. The resonant inductor limits dI/dt of the diode current, so its reverse-recovery problem and the associated switching losses are significantly reduced. After the current of the resonant inductor, I_{L_x}, reaches I_a at t_2, diode D_{ap} turns off and switch S_{ap} conducts the surplus current beyond the current I_a until enough initial resonant energy is stored in the resonant inductor. This is why S_{ap} should be on when D_{ap} turns off. S_{ap} is turned off at t_3, the ensuing resonance between the resonant inductor, L_x, and the capacitor, C_S, paralleled to S_{an} brings the voltage across this switch to zero ($t_3 - t_4$). During $t_3 - t_4$, I_{L_x} varies in resonant fashion (sinusoidal) and the resonant current is added to the current of resonant inductor at t_3.

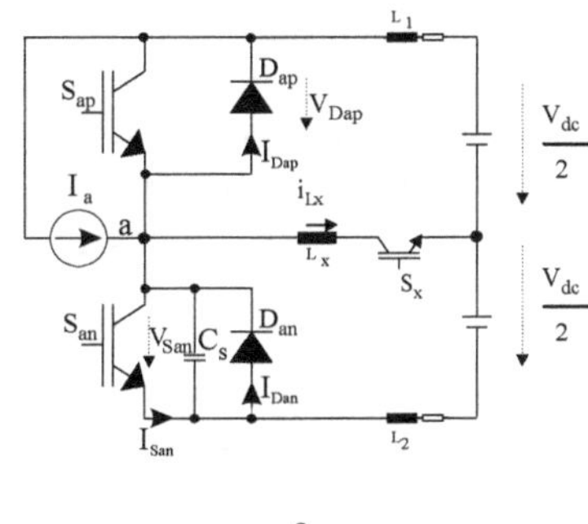

a

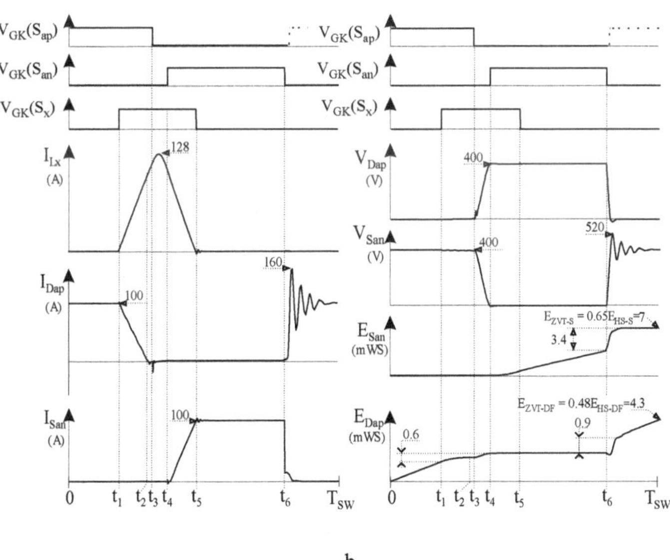

b

Fig. 2.5: Zero Voltage Switching (ZVS): a) circuit diagram b) operating waveforms.

The initial resonant energy is required for covering the conduction losses of resonant branch during the resonance between the inductor, L_x, and the capacitor, C_S, and for providing some surplus current when this resonance ends. The free-wheeling of this surplus current by the antiparallel diode, D_{an}, creates the zero voltage conditions for turn-on of the switch S_{an}. After S_{an} is turned on, the current of the resonant inductor is linearly reduced to zero because the voltage across it (-$V_{dc}/2$) is now reversed ($t_4 - t_5$). At t_5, I_{L_x} reaches zero and S_x is turned off.

When the switch S_{an} is turned off at t_6, due to the resonant capacitor C_S across the switch, its voltage $V_{S_{an}}$ cannot rise as quickly as in case of hard-switching. This helps to reduce the turn-off loss of the switch. With adding an additional capacitance across the switch, the major part of the turn-off loss appears as the capacitive turn-on loss. The capacitive turn-on loss is then eliminated by the ZVS technique. However, after the sudden removal of the gate drive signal and the subsequent current fall, the ensuing current pump remains and is even larger than the tail current in the hard-switching case. Adding a larger capacitor is certainly helpful to further slow down the voltage rise and reduce the turn-off loss, but, this may not be practical from the circuit point of view, for a larger capacitor introduces higher circulating energy and additional conduction loss in the auxiliary branch.

In summary, the ZVS operation eliminates the IGBT turn-on loss and solves the intricate diode reverse-recovery problem. Moreover, it can also reduce the turn-off loss to some extent.

2.2.3 Zero Current Switching

ZCS technique is usually employed to alleviate turn-off loss of devices. Figure 2.6(a) shows the circuit of Fig. 2.4 with a ZCS scheme. The resonant branch (S_x, D_x, L_x and C_s) is added to the hard switching circuit to achieve ZCS turn-off for the power switch S_{an} [35].

Figure 2.6(b) shows the operating waveforms of the circuit. Assume that the circuit is in the same initial state as in the case of hard switching and that the resonant capacitor C_S is charged to an initial voltage $V_{C_S}(t_1) = V_o$. The power switch S_{an} is turned on at t_1. With the turn-on of the switch the resonant

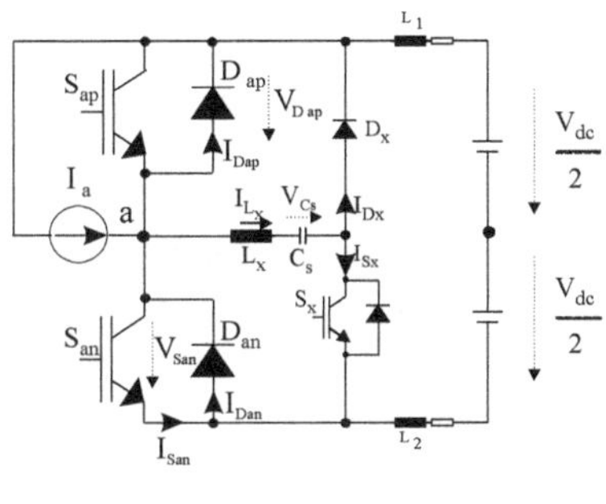

a

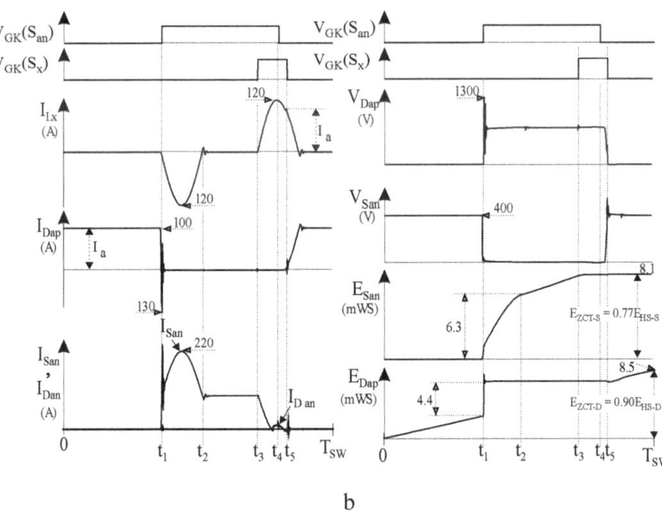

b

Fig. 2.6: Zero Current Switching (ZCS): a) circuit diagram b) operating waveforms.

branch is activated through the antiparallel diode of the auxiliary switch S_x. The resonant branch oscillates a half-cycle, thereby the voltage across the resonant capacitor V_{C_S} changes its polarity. The dynamic behavior of the switch S_{an} and diode D_{ap} during the turn-on transition $(t_1 - t_2)$ is essentially similar to that described for the hard-switching circuit except that in the circuit of Fig. 2.6(a) the resonant current caused by the polarity changing of the resonant capacitor C_S is superimposed on the switch current resulting in some additional conduction loss in S_{an}. All the aforementioned problems related to turn-on transition such as turn-on loss, high dV/dt and high dI/dt resulting in the reverse recovery problem of the diode D_{ap} are left in this switching technique unsolved.

Before the turn-off time of the switch S_{an}, the auxiliary branch is activated at t_3 by turning on the auxiliary switch S_x. Because the voltage across the resonant capacitor V_{C_S} is now negative, the resonant branch starts resonating and diverts the current of the switch $I_{S_{an}}$ to the resonant branch. After the switch current reaches zero, the antiparallel diode D_{an} conducts the surplus current. This allows a hold-off time for turning off the switch S_{an} under zero current conditions which happens at t_4. At t_5, the resonant current becomes smaller than I_a causing the diode D_{an} to turn off and the voltage across the switch S_{an} to rise. After t_5, the auxiliary switch S_x is turned off and the current remaining in the resonant branch is transferred to the clamp diode D_x. During the half-cycle oscillation of the resonant branch, the voltage across C_S changes again its polarity. It reaches again the initial value V_o at the end of half-cycle oscillation when the resonant current I_{L_x} reaches zero and the clamp diode D_x turns off.

This technique only eliminates the turn-off loss. Therefore it is worthy to use for slow devices whose major losses occur at turn-off instant.

2.2.4 Comparison of Total Switching Losses for Different Switching Techniques

The waveforms shown in sections 2.2.1 to 2.2.3 are derived from the simulation of the circuits with simulator SABER [38] using accurate physics-based dynamic models of IGBT and power diode considering dynamic behaviours of semiconductor devices such as reverse and forward recovery effects of power diodes and also the turn-off tail current of IGBTs. These models can be used to describe the dynamic behavior and power dissipation of IGBTs and Power Diodes in user-defined circuits [39]-[41]. The circuits are simulated with $V_{dc} = 400V$ and $I_a = 100A$. The parameters of models are reasonably set, so that they can easily withstand the voltage and current stresses of hard switching conditions. The most important parameters are listed as follows:

Switch S_{an} (IGBT):

Device active area, $A = 2.5\,cm^2$,

Base width, $W = 0.017\,cm$,

High-level injection lifetime $\tau_{H0} = 7\,\mu s$,

Freewheeling diode D_{ap}:

Device active area, $A = 3.17\,cm^2$,

Drift region length, $W = 0.0235\,cm$,

High-level injection lifetime $\tau_{H0} = 1.5\,\mu s$.

The simulations are carried out under the same switch current at turn-on and turn-off times (100A) and under the same gate resistance for all cases. The losses of the circuits are calculated from the integral of product of voltage across the device times current through the device. The interval T_{sw} is also the same for all cases. The results of loss calculation are summarized in table 2.1. In this table the first four columns show turn-on and turn-off energy losses for the IGBT power switch S_{an} and power diode D_{ap}. The next two columns are the device losses including the conduction loss for the interval T_{sw}. The sum of losses of the switch and diode is given in the last column. Gate circuit energy losses are ignored in this table.

As can be seen on the table 2.1, the turn-on losses (including the diode reverse-recovery losses) are higher than the turn-off loss. This itself suggests the effectiveness of ZVS operation in reduction of switching losses. The results show a 33% reduction in the switch total losses and a 53% reduction in the diode loss using ZVS technique. This results in a 43% reduction in the sum of the switch and diode losses. This technique also relieves the turn-on current stress of the switch caused by the reverse recovery problem of the diode. The ZCS technique also minimizes the turn-off voltage stress on the switch and causes a reduction of 23%, 8% and 16% in the switch total losses, the diode total losses and the sum

Device Loss Energy for Interval T_{sw} in mWS							
Switching	*Turn-on Loss*		*Turn-off Loss*		*Device Losses*		*Sum of*
Technique	$E_{S_{an}}$	$E_{D_{ap}}$	$E_{S_{an}}$	$E_{D_{ap}}$	$E_{S_{an}}$	$E_{D_{ap}}$	*Losses*
HS	1.9	4.4	4.5	1.2	10.5	9.2	19.7
ZVS	0.0	0.6	3.4	0.9	7.0	4.3	11.3
ZCS	6.3	4.4	0.0	0.0	8.1	8.5	16.6

Table 2.1: Energy Losses for the interval T_{sw} in mWS.

of the switch and diode losses, respectively.

The zero voltage soft switching technique seems to be more effective. The attainable advantages of the technique are reduction of the switching losses, reduction of voltage spike on switches at turn-off, limitation of di/dt and dv/dt and from there elimination of the problems related to reverse-recovery-effect of diodes (see also figures 2.4, 2.5 and 2.6).

2.3 AC-Side Soft Commutated PWM Three Phase Converters

In the ASSC PWM converters an auxiliary resonant circuit is used at the AC-side to create soft switching conditions for all converter switches. Fig.2.7 shows the general topology of an AC-side soft commutated PWM converter in which the auxiliary circuit can be one of the circuits introduced in the following sections. During the switching transition, the auxiliary circuit is activated to discharge the parasitic and additionally paralleled capacitances across the on-going switches. After the transition is completed, the auxiliary network becomes inactive, and the converter remains functionally identical to its hard-switched PWM counterpart. Since the auxiliary network processes only a small fraction of the total converter power, the AC-side auxiliary circuit is well-suited for high power applications.

The switches of an AC-side auxiliary circuit have to withstand either full DC link voltage or even half of it depending on the topology of auxiliary circuit. The

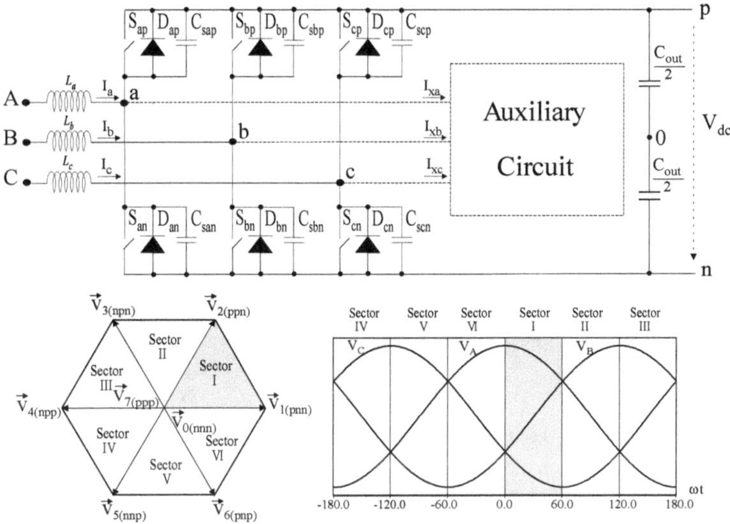

Fig. 2.7: ZVT three-phase converter, three-phase voltages and space voltage vectors.

maximum current of the auxiliary switches can reach a value twice as large as the amplitude of phase current, but their average current remains small because the auxiliary switches conduct only during the switching transition which is very short compared to the switching cycle.

The basic purposes for employing an AC-side auxiliary circuit are:

- Providing the ZVS conditions for all main switches of the converter,
- Minimizing the entire losses of the converter including the losses of the auxiliary circuit.

In this regard a few important points should be considered in evaluating the different AC-side auxiliary circuits:

- The space vector modulation and the quality of AC waveforms should not be possibly distorted by the auxiliary circuit operation.

- The capability of space vector modulation should not be possibly limited due to the requirements of the auxiliary circuit operation.
- The complexity of auxiliary circuit should be as low as possible.
- The switching control timing should be as simple as possible.

This section discusses the operation principles of the ASSC PWM converters along with their modified SVM with help of the results obtained from the simulation of the proposed circuits using simple models of the semiconductor devices without considering their dynamic behaviour. This section also evaluates the converters from the point of views mentioned above and describes how they can be improved.

For simplicity of discussion, the converters are assumed to operate in sector I of the voltage hexagon as shown in gray in Fig.2.7, for operation of the converters is symmetrical in every 60° interval of SVM. The node voltage V_a, V_b and V_c are measured with respect to the center point of the DC-side capacitance and the direction of the phase currents into the converter is assumed to be positive. In Fig.2.7 the voltage waveform of nodes A, B and C are measured with respect to neutral point of the AC-side (not shown) and $\vec{V}_{0(nnn)}$ to $\vec{V}_{7(ppp)}$ are the eight different voltage vectors produced by the eight different switching states $S_{0(nnn)}$ to $S_{7(ppp)}$ shown in Fig. 2.3(a).

2.3.1 AC-Side Half-Wave Auxiliary Resonant Commutated Converter

The circuit presented in [36] is the only example of a so-called AC-Side Half-Wave Auxiliary Resonant Commutated (ASC-HW-ARC) converter with one switch in the auxiliary circuit. Fig. 2.8 shows this circuit. This circuit is advantageous from the point of view of number of active devices in the auxiliary circuit. The auxiliary resonant circuit of this converter provides the ZVS conditions for the on-going switches after a half-cycle oscillation. For this reason the category of this converter is named AC-side half-wave.

The auxiliary circuit is only activated for diode-to-switch current commutations. The switch-to-diode commutation of the phase currents is a charging/discharging process which takes place automatically by opening the switch.

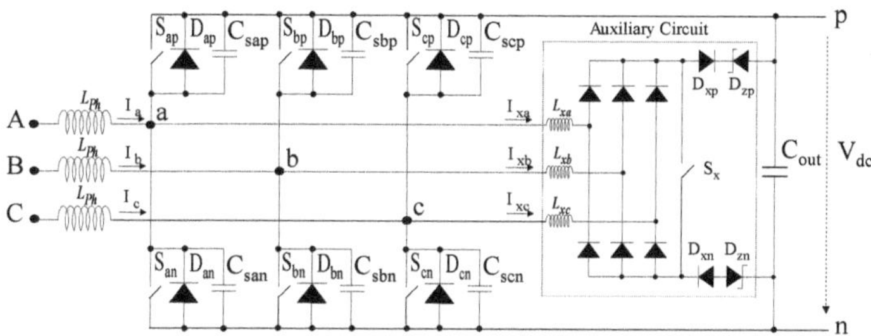

Fig. 2.8: ASC-HW-ARC PWM converter with one switch at the auxiliary circuit.

For example, suppose the switch S_{an} is on and conducts the phase current, I_a, assumed to be positive. When this switch is turned off, the phase current charges the capacitor C_{san} across the switch and discharges the capacitor C_{sap} across the switch S_{ap} linearly. The node voltage V_a changes linearly from the negative DC rail voltage to the positive DC rail voltage. When the node voltage reaches the positive DC rail voltage, diode D_{ap} starts contacting and the commutation ends successfully without the activation of the auxiliary circuit. If the node capacitances are selected properly, there is no need to activate the auxiliary circuit for a switch-to-diode current commutation unless the phase current is very small and the commutation of phase current lasts an unacceptable time that is far too long This kind of commutation is also called *self-commutation* in this work and happens softly when the switch conducting the phase current is turned off. The auxiliary circuit is then required for diode-to-switch current commutations when the phase current should be commutated from a conducting diode to a on-going switches located at the same leg. In the example considered above this is when the phase current I_a has to be commutated from diode D_{ap} to switch S_{an}.

2.3.1.1 Auxiliary Circuit Operation

The above discussion reveals that if the diode-to-switch current commutation of all three phases are synchronized, all three phase currents can be commutate with only one Auxiliary Circuit Action (ACA). For this purpose the auxiliary circuit always has to be activated at the end of duration of switching state in which diodes are conducting the phase currents. It will be shown in section 2.3.1.2 that at any load condition there exists one switching state which satisfies this condition and has three conducting diodes. This switching state, which is necessary for operation of the auxiliary circuit of this converter, may be even an undesired switching state.

In the rectifying operation when, for example, I_a, I_b and I_c are positive, negative and negative respectively, the desired switching state $S_{1(pnn)}$ has three conducting diodes whereas in the inverting mode when, for example, I_a, I_b and I_c are negative, positive and positive respectively, the undesired switching state $S_{4(npp)}$ has three conducting diodes. This subject will be also discussed later in section 2.3.1.2.

In following it is assumed that at a given instant I_a, I_b and I_c are positive, negative and negative respectively and the circuit is in the switching state $S_{1(pnn)}$ producing the voltage vector $\vec{V}_{1(pnn)}$ at the AC side. Diodes D_{ap}, D_{bn} and D_{cn} conduct the phase currents and switches S_{ap}, S_{bn} and S_{cn} are also on[5]. The phase currents should be commutated from diodes D_{ap}, D_{bn} and D_{cn} to switches S_{an}, S_{bp} and S_{cp} with only one ACA starting at the end of duration of switching state $S_{1(pnn)}$. Fig. 2.9(a) shows the converter waveforms during this ACA. Fig. 2.9(b) shows the corresponding equivalent topological states of the circuit as they change.

$t_1 - t_3$: A) Charging and B) boost charging states

The switching state S_{pnn} applies the voltage V_{dc} at the input of the auxiliary circuit between the nodes a and b, and the nodes a and c. The auxiliary circuit action (ACA) starts at t_1 by closing the switch S_x, which brings the auxiliary circuit into the charging state as shown in Fig. 2.9(b-A). The auxiliary inductor currents I_{xa}, I_{xb} and I_{xc} start to increase linearly in the same direction of phase currents I_a, I_b and I_c causing the current of the diodes D_{ap}, D_{bn} and D_{cn} to decrease slowly. When the auxiliary inductor currents exceed the phase currents, D_{ap}, D_{bn} and D_{cn} turn off softly. After the diodes turn off, their antiparallel

[5] Due to the operation of the auxiliary circuit these switches should be on before ACA begins.

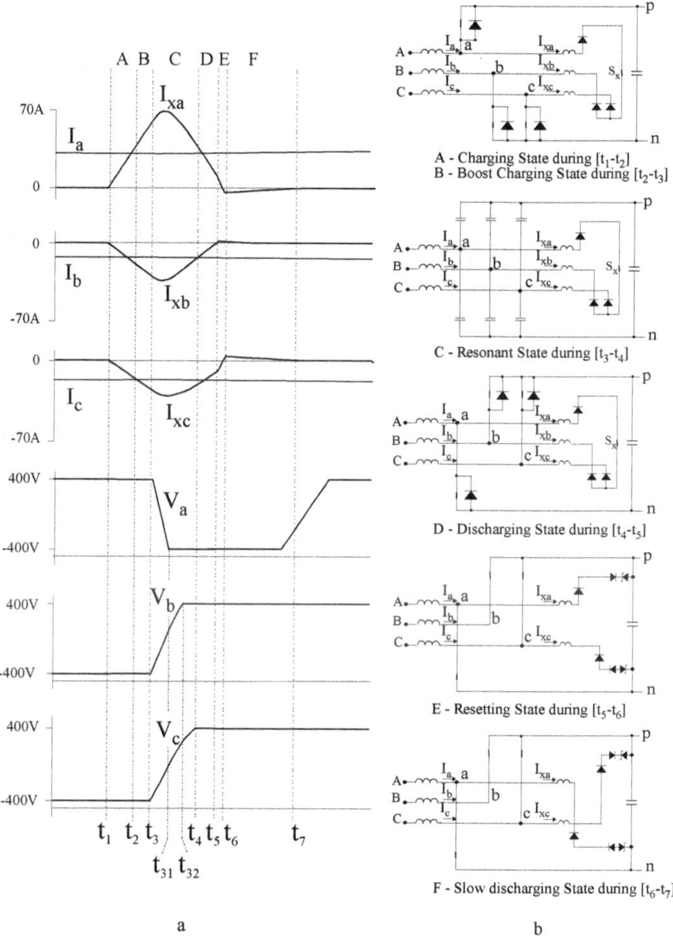

Fig. 2.9: a) Operation waveforms of the ACS-HW auxiliary circuit and b) its topological sequence for an ACA.

switches S_{an}, S_{bp} and S_{cp} conduct the surplus of the auxiliary currents. The auxiliary inductors necessarily continue charging until enough energy is stored in them ($t_2 - t_3$ in Fig. 2.9(a)). This additional energy is named boost energy[6]. At the end of the boost charging, t_3, S_{ap}, S_{bn} and S_{cn} are turned off simultaneously. The capacitances C_{sap}, C_{sbn} and C_{scn} do not allow the voltage across the switches to rise rapidly, consequently, the turn-off action of these switches under the surplus auxiliary currents cannot cause much turn-off losses.

It is worth noting that only the switching states producing the non-zero voltage vectors can have three conducting diodes, since conduction of three diodes in a zero switching state means the same current direction for all three phases which is not true. Thus, the switching state with three conducting diodes always ensures the existence of sufficient voltage at the auxiliary circuit in the charging state. It also ensures the correct direction of the auxiliary currents. Considering the currents coming into/out from the node a, b or c, the auxiliary currents always start to increase in the same direction as the current of conducting diodes are flowing. i.e. coming into the node or coming out from the node. In this way the increase of the auxiliary currents always results in decrease of the current of conducting diodes and deviation of the phase currents from the conducting diodes into the auxiliary circuit.

$t_3 - t_4$: C) Resonant states

With the turn-off of the switches S_{ap}, S_{bn} and S_{cn}, their parallel capacitances take effect and the circuit is transferred into the resonant state as shown in Fig. 2.9(b-C). The node voltages V_a, V_b and V_c change in a resonant fashion towards the opposite DC rail from where they started. The node voltage V_a reaches the negative DC rail faster than the voltage of nodes b and c, for the initial resonant energy of L_{xa} (resp. its current) at the beginning of resonant state is higher than that of other auxiliary inductors[7]. At t_{31}, diode D_{an} starts conducting and provides the ZVS conditions for S_{an}. The resonant state of nodes b and c continues until the node voltages V_b and V_c reach the opposite DC rail voltage. This happens for V_b and V_c at t_{32} and t_4 respectively. Because the diode

[6] The boost energy is necessary for complete charging the capacitances C_{sap}, C_{sbn} and C_{scn} and discharging the capacitances C_{san}, C_{sbp} and C_{scp} in the resonant stage. It will be shown in chapter 3 that without enough boost energy the ZVS conditions cannot be completely achieved for all phases, due to the on-state losses of the auxiliary switch and diodes, and more important, due to functionally changing of the topology of the resonant network during the resonant stage (see 3.1.1).

[7] Because $I_{xa} = I_{xb} + I_{xc}$.

D_{an}, D_{bp} and D_{cp} start conducting at different times, the topology of resonant circuit changes functionally during the resonant state. If enough initial energy have not been stored in the auxiliary inductors at the end of the boost charging state, the ZVS conditions cannot be completely achieved[8)] for phases b and c resulting in an increase in the switching losses of these phases. At t_4, the end of the resonant state, when all three diodes D_{an}, D_{bp} and D_{cp} are conducting, the switches S_{an}, S_{bp} and S_{cp} can be turned on together under the zero voltage conditions. As can be seen from the waveform of Fig. 2.9(a) the resonant circuit oscillates a half-cycle. The group of this circuit is therefore named AC-side half-wave resonant circuits compared to the circuits introduced in sections 2.3.3 and 2.3.4.

$t_4 - t_5$: D) Discharging state

With the turn-on of S_{an}, S_{bp} and S_{cp}, V_a, V_b and V_c are clamped to the negative, positive and positive DC rail respectively. This reverses the input voltage of the auxiliary circuit and also the voltage across the auxiliary inductors, thereby the auxiliary currents decrease linearly and the major energy remaining in the the auxiliary inductors is discharged to the AC-side. When I_{xb} reaches zero at t_5, the magnitude of I_{xa} is also small and S_x can be turned off at a small turn-off current.

$t_5 - t_6$: E) Resetting state

Opening of S_x leads to a resonance with short duration between the auxiliary inductors and the parasitic capacitance of S_x. After the parasitic capacitance of the switch S_x has been charged, the circuit is transferred to the resetting state as shown in Fig. 2.9(b-E). The auxiliary currents I_{xa} and I_{xc} decrease quickly because full DC voltage V_{dc} now appears across each auxiliary inductor.

$t_6 - t_7$: F) Slow discharging state

At the end of the resetting state, another short resonance occurs between the auxiliary inductors and the parasitic capacitances of the auxiliary bridge diodes, which reverses the currents I_{xa} and I_{xc}. As a result the circuit goes to the slow discharging state as shown in Fig. 2.9(b-F), where the energy remaining in the auxiliary inductors is slowly dissipated in the series connection of a main switch,

---

[8)] See also section 3.1.1.

two rectifier diodes and a zener diode. At t_7, I_{xa} and I_{xc} reach zero and the auxiliary circuit action ends.

2.3.1.2 Switching Sequence and SVM Requirement

The above example shows that the auxiliary circuit is able to provide zero voltage turn-on conditions for the on-going switches and soft turn-off conditions for the diodes turning off, when it is activated in a switching state with three conducting diodes.

Generally, the switching state with the phase voltage polarities similar to the phase current directions has always three conducting diode. In the example of last section the phase currents I_a, I_b and I_c were positive, negative and negative respectively and the polarities of node voltages V_a, V_b and V_c in the switching state $S_{1(pnn)}$ were too. Since every one of the six possible combinations of the phase current directions (for example positive, positive and negative) corresponds to one of the six non-zero switching states, which has the similar polarities for the phase voltages ($S_{2(ppn)}$ in this example), it is always possible to find a switching state having three conducting diodes at any load condition.

The switching state with three conducting diodes is a desired switching state as long as the converter operates in the rectifier mode of operation. In this mode the directions of phase currents are similar to the polarities of phase voltages in one of the desired non-zero switching states. Considering the reference voltage vector $\vec{V}_{ref}$ in sector I of voltage hexagon[9)], the switching state with three conducting diodes is $S_{1(pnn)}$ or $S_{2(ppn)}$ when the phase currents I_a, I_b and I_c are positive, negative and negative or positive, positive and negative respectively. The first is the case of example of last section. In the inverter and transient modes of operation the switching state having three conducting diodes is an undesired switching state. In the inverter mode of operation the undesired switching state $S_{4(npp)}$ or $S_{5(nnp)}$ has three conducting diodes when the phase currents I_a, I_b and I_c are negative, positive and positive or negative, negative and positive respectively.

It is shown that an ACA starts from the switching state with three conducting diodes and ends in the switching state with three conducting switches. From

[9)] The discussion is analogous for other sectors.

this state, any two switching states producing two adjacent voltage vectors can be attained by opening the switches. By selecting a proper switching sequence, it is possible to synchronize the diode-to-switch current commutation of all three phases and hence to activate the auxiliary switch only once per switching cycle. Now, if ACA is placed at the beginning of switching sequence, the auxiliary switch can be switched with the same switching frequency as the main switches are controlled.

The voltage vectors before and after an ACA are always complementary, because all three node voltages change their polarity during the ACA. Since two complementary vectors are never adjacent, an undesired voltage vector is always produced due to ACA. In the example of last section, the auxiliary circuit is activated when the diodes D_{ap}, D_{bn} and D_{cn} are conducting and the circuit produces the voltage vector $\vec{V}_{1(pnn)}$. After the ACA, the circuit is brought to the switching state with three conducting switches S_{ap}, S_{bn} and S_{cn} producing $\vec{V}_{4(npp)}$, the complement of $\vec{V}_{1(pnn)}$. Production of $\vec{V}_{4(npp)}$ is undesired, but the circuit necessarily produces it to reverse the voltage at the auxiliary circuit for discharging the auxiliary inductors. After finishing the discharging state, the circuit can be switched to one of the desired switching states. When the switching sequence $S_{1(pnn)} - ACA - S_{4(npp)} \rightarrow S_{7(ppp)} \rightarrow S_{2(ppn)}$ is followed in sector I, other transitions are accomplished by opening the switch, one at a time. The effect of the undesired voltage vector $\vec{V}_{4(npp)}$ in the synthesis of $\vec{V}_{ref}$ can be canceled by extending the duty cycle of its complementary desired voltage vector $\vec{V}_{1(pnn)}$ by the same amount. This generates an effective zero voltage with the duty cycle twice as much as the duty cycle of the undesired vector T_c. Thus the duty cycle of the zero voltage vector $\vec{V}_{7(ppp)}$ has to be reduced by $(2\,T_c)$ as shown in Fig. 2.10(b top). Therefore, instead of (2.4), reference voltage vector $\vec{V}_{ref}$ is now realized according to:

$$V_{ref} = (t_1 + t_c)\,\vec{V}_{1(pnn)} + t_c\,\vec{V}_{4(npp)} + t_2\,\vec{V}_{2(ppn)} + (t_7 - 2\,t_c)\,\vec{V}_{7(ppp)}, \tag{2.5}$$

where t_1, t_2 and t_7 are the normalized duty cycles. $T_c = t_c\,T_S$ is the minimum duty cycle of the undesired voltage vector and should be large enough for a complete ZVT process.

In the normal rectifying operation corresponding to the previous example, the undesired switching state is used for the discharging state, therefore this

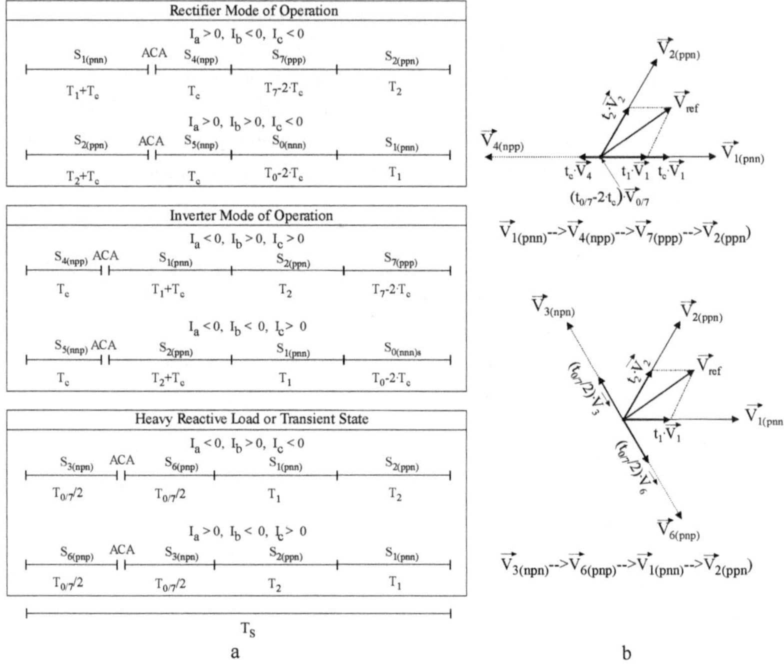

Fig. 2.10: Modified SVM for the ACS-HW-ARC converter.

switching state needs to be switched only for a short time slightly larger than the discharging duration.

In the normal inverting operation, the switching state with three conducting diodes produces an undesired voltage vector which is complement of one of desired voltage vectors. The auxiliary circuit is activated at the end of this state. Since this state is used for the charging state of ACA, its duration should be slightly longer than the charging duration. ACA then transfers the circuit to a desired switching state with three conducting switches. The remaining necessary switching states are achieved by opening the switches, one at a time. After the final switch is opened, the circuit is again in the switching state with three conducting diodes. Fig. 2.10(a) shows switching sequences for different modes of operation.

In the transient cases, or in the case of the phase shifts between the phase current and phase voltage being more than 30° lagging or leading, the switching state with three conducting diodes produces none of the desired voltage vectors nor their complements. However, the intended reference voltage vector $\vec{V}_{ref}$ can still be realized with only one auxiliary circuit action by eliminating the zero voltage vector from the SVM sequence. Fig. 2.10(b bottom) shows the synthesis of $\vec{V}_{ref}$ when the switching state with three conducting diodes produces the undesired voltage vector $\vec{V}_{3(npn)}$. After ACA, the circuit is brought to its complementary state producing another undesired voltage vector, $\vec{V}_{6(pnp)}$. If these vectors are applied for the same duration, they will cancel one another and actually produce the zero vector effect. As shown in Fig. 2.10(b bottom), the intended vector $\vec{V}_{ref}$ can be synthesized now as follows:

$$\vec{V}_{ref} = t_1 \vec{V}_1 + t_2 \vec{V}_2 + \frac{t_{0/7}}{2} \vec{V}_3 + \frac{t_{0/7}}{2} \vec{V}_6. \tag{2.6}$$

Although the influence of the undesired voltage vector on the synthesis of $\vec{V}_{ref}$ will be canceled, production of this voltage vector results in increase of the amplitude of high order harmonics and worsening of the waveform quality of AC currents.

2.3.1.3 Problems and Solutions

The circuit is advantageous from the view point of having only one active device in its auxiliary circuit. Also, the auxiliary switch operates under soft switching conditions. But the circuit has to produce an undesired voltage vector for the auxiliary circuit operation, because no voltage is available for charging/discharging of the auxiliary inductors when the main circuit produces zero voltage vectors. It results in following serious problems:

1. Production of an undesired voltage vector results in increase of the amplitude of high order harmonics of AC quantities leading to worsening of their waveform quality.

2. Duration of the undesired voltage vector and expansion of the duration of its complementary desired voltage vector shorten the effective usable duration of sampling cycle for synthesizing $\vec{V}_{ref}$. This results in a reduction

in the linearity range of the space vector modulation and reduction of DC bus utilization.

3. Additional switching actions are necessary for switching between the undesired and desired switching states. Since switch turn-off is not completely lossless, it further increases the switching turn-off losses.

4. In case of a switch-to-diode current commutation with small magnitude of phase current, this auxiliary circuit is not able to accelerate the current commutation. In this case, the circuit either enjoys the soft switching conditions at the expense of a long self-commutation process, resulting in a reduction of DC bus utilization, a problem to be discussed in chapter 5, or soft switching conditions have to be abandoned which results in further increase of the switching losses because of parallel capacitances.

5. Considering the problem of undesired switching state from the design point of view, the components of the auxiliary circuit, or even the components of the main circuit should be overdesigned to ensure the commutation capability of circuit for commutating the current maximum increased excessively due to the switching of the undesired voltage vector.

To solve these problems and to get rid of the undesired voltage vector, the auxiliary circuit should be modified so that enough voltage can be applied across the auxiliary inductors even at zero voltage vectors. The solution would be creation of a midpoint at the DC-side and connection of the auxiliary inductors to this point. Furthermore, independent commutation of the phase currents seems to be more desirable, because there is no need to commutate all three phase currents for switching between any two desired switching states. To minimize the losses and the number of switching actions, it is also preferable that only the auxiliary components related to the phase currents, which should be commutated, take part in commutation action.

For this purpose an auxiliary circuit with three separate auxiliary branches, the so-called pole commutating circuit, is necessary. The pole commutation concept offers many advantages at the expense of six active devices in the auxiliary circuit which is the subject of the next section.

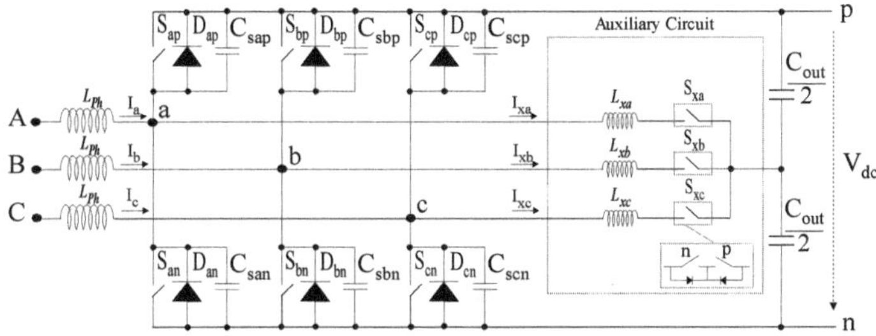

Fig. 2.11: The auxiliary resonant commutated pole converter.

2.3.2 AC-Side Half-Wave Auxiliary Resonant Commutated Pole Converter

The Auxiliary Resonant Commutated Pole (ARCP) converter [31] was the first topology which utilized the concept of AC-side pole commutation. Figure 2.11 shows this circuit. Recalling the auxiliary circuit described in the previous section, the freedom degrees[10)] of the modulation were utilized to create a certain SVM scheme which meets the requirements of the auxiliary circuit operation because the auxiliary circuit treats the switching problem of all three phases together. On the contrary, the auxiliary circuit of ARCP converter operates independently in every phase, so the converter can be controlled with any SVM scheme. This is a very advantageous ability because the freedom degrees of space vector modulation can then be utilized for optimizing other criteria such as losses (see chapter 4).

Figure 2.12 illustrates the slightly modified ARCP circuit, which is named the AC-Side Half-Wave Auxiliary Resonant Commutated Pole (ACS-HW-ARCP) converter in this work. This circuit offers additional advantages in simultaneous commutation of two phase currents, which are described in the next section.

10) A subject to be discussed in chapter 4.

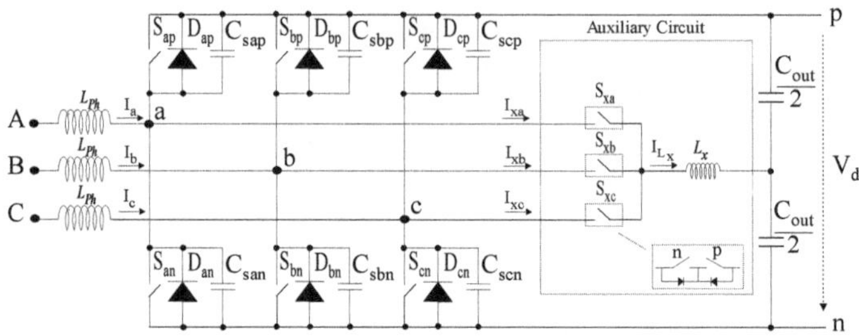

Fig. 2.12: The AC-side half-wave auxiliary resonant commutated pole converter.

The principles of the auxiliary circuit operation in each phase of this converter is similar to that described in section 2.3.1.1 and consists of charging, resonant and discharging stages. The auxiliary circuit action will be briefly explained in the rectifier and inverter mode of operation. The circuit waveforms for both modes of operation are shown in Fig. 2.13. It should be noted that the voltage across the auxiliary inductor L_x during the charging and discharging stages is half of the DC bus voltage because it is connected to the center point of the DC bus capacitances.

2.3.2.1 Auxiliary Circuit Operation

Consider again the same initial conditions as in the example of ACS-HW-ARC circuit, i. e. the circuit is in the switching state S_{pnn} and produces the voltage vector $\vec{V}_{1(pnn)}$ at the AC-side, the switches S_{ap}, S_{bn} and S_{cn} are on. Phase currents I_a, I_b and I_c flowing through the diodes D_{ap}, D_{bn} and D_{cn} are positive, negative and negative respectively. The circuit should still produce $\vec{V}_{2(ppn)}$ and a zero voltage vector for synthesizing $\vec{V}_{ref}$ located in voltage sector I. Suppose that the circuit should be transferred to the switching state $S_{7(ppp)}$ for the time being, i. e. the phase currents I_b and I_c have to be commutated from the diodes D_{bn} and D_{cn} to the switches S_{bp} and S_{cp}.

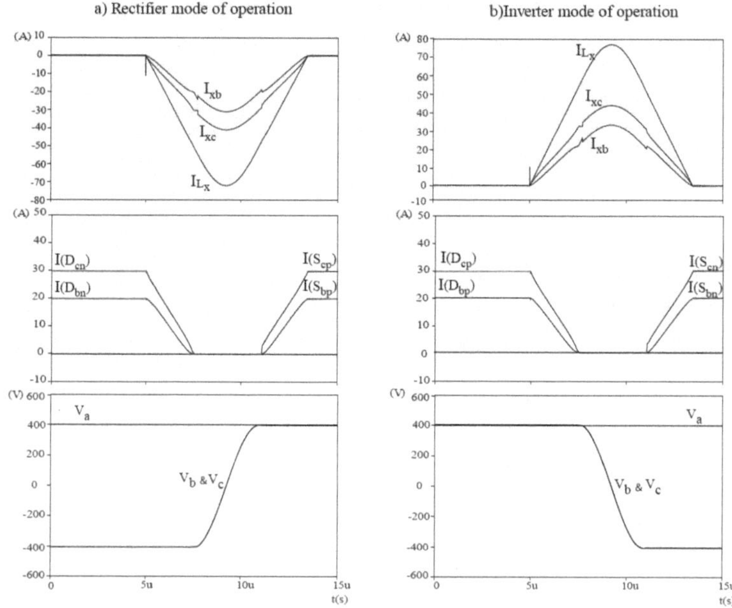

Fig. 2.13: Waveforms of the ACS-HW auxiliary resonant commutated pole circuit in rectifier and inverter modes of operation.

The auxiliary circuit action (ACA) starts when the auxiliary switches $S_{xb(n)}$ and $S_{xc(n)}$ are turned on. Half of the DC bus voltage ($V_{dc}/2$) is applied across L_x. The principle of operation in the charging state is the same as described in 2.3.1.1 except that the auxiliary inductor, L_x, has to charge until the magnitude of its current, $|I_{L_x}|$, reaches the value $(|I_b| + |I_c| + I_{boost})$. I_{boost} determines the required boost energy for covering the conduction loss of the resonant circuit, for compensating the deviation of DC-side center point voltage from zero and for providing a ZVS hold-on time for the on-going switches, a subject to be discussed in section 2.3.2.4 and to be analysed in detail in chapter 3. Without enough boost energy the ZVS conditions cannot be completely achieved.

With the increase of the auxiliary inductor current, the current of the diodes D_{bn} and D_{cn} decreases. When the auxiliary inductor current I_{L_x} reach the value

$|I_b + I_c|$, D_{bn} and D_{cn} turn off simultaneously and the switches S_{bn} and S_{cn} start conducting the surplus currents. Note, that because of the different voltage drop across the diode D_{bn} and D_{cn}, which is caused by differing I_b and I_c, the auxiliary inductor current is not equally divided between the phases b and c. After enough boost energy is stored in the auxiliary inductor, both switches S_{bn} and S_{cn} are simultaneously turned off resulting in an equal division of boost current between the phases b and c. After S_{bn} and S_{cn} turn off, the capacitances at the nodes b and c take effect and the resonant state begins. Equal division of I_{boost} between the phases b and c leads to the same variation of the node voltages V_b and V_c during the resonant state. V_b and V_c reach the positive DC rail voltage at the same time. When D_{bp} and D_{cp} start conducting, S_{bp} and S_{cp} can be turned on under zero voltage conditions simultaneously. The voltage across the auxiliary inductor is reversed during the resonant state causing the discharging stage to start automatically after V_b and V_c reach the positive DC rail voltage. It is worth mentioning that the discharging state starts although the converter produces a zero voltage vector ($\vec{V}_{7(ppp)}$) at the AC-side. When the current I_{L_x} becomes zero at the end of ACA, $S_{xb(n)}$ and $S_{xc(n)}$ are turned off lossless under zero current conditions.

In the inverter mode of the above example, phase currents I_a, I_b and I_c are now negative, positive and positive respectively. The auxiliary branches of phases b and c have to be activated for the transition $S_{7(ppp)} \rightarrow S_{1(pnn)}$.

The ACA begins in this case with turning on of $S_{xb(p)}$ and $S_{xc(p)}$. The auxiliary current I_{L_x} increases in the opposite direction of the rectifier mode. The transition occurs in the same manner as explained for the rectifier mode. The operation waveforms of the inverter mode of operation are shown in Fig. 2.13(b).

Considering $\vec{V}_{ref}$ in sector I, the voltage vectors synthesizing $\vec{V}_{ref}$ are $\vec{V}_{1(pnn)}$, $\vec{V}_{2(ppn)}$ and a zero vector. With the sequence $ACA_{b,c}$[11] -$S_{7(ppp)} \rightarrow S_{2(ppn)} \rightarrow S_{1(pnn)}$ in the rectifier mode, the auxiliary switches of phases b and c are activated for the transition $S_{1(pnn)} \rightarrow S_{7(ppp)}$. The other switching states can be attained from $S_{7(ppp)}$ by opening the switches S_{bp} and S_{cp}, one at a time. With placing the auxiliary circuit action (ACA) at the beginning of the switching cycle, the auxiliary switches can be switched with the same switching frequency as

[11] $ACA_{b,c}$ means that the auxiliary branches related to phases b and c take part in ACA. ACA_b means that only the auxiliary branch related to phases b takes part in ACA.

the main switches are controlled.

Since soft commutation of phase currents occurs independently in every phase, from the point of view of circuit operation it is not important which switching sequence is used. One can use the sequence $S_{1(pnn)}$-ACA_b-$S_{2(ppn)}$-ACA_c-$S_{7(ppp)}$ in which $S_{xb(n)}$ is activated for the first transition and $S_{xc(n)}$ for the second one. Anyway, the resulting losses remain the same assuming the phase currents to be constant within the switching cycle. Although the auxiliary switches have no longer a constant switching frequency using this sequence, it can be advantageous when a switching pattern like 127721 is used. Other transitions of this pattern are the so-called self-commutation transition and take place by opening the respective switches. The switching losses will be considerably reduced using this pattern because the switching frequency f_{sw} of this pattern is half of the sampling frequency f_s (see chapter 4).

It must be mentioned that in the switching sequence mentioned above ACA_c can start, after the resonant stage of ACA_b is finished. An earlier start of ACA_c results in incomplete ZVS conditions for switch S_{bp}. This is because of this fact that the auxiliary inductor L_x is a common component for the three auxiliary branches of the ACS-HW-ARCP circuit. It is not the case of conventional ARCP circuit shown in Fig. 2.11, since the auxiliary branches of this circuit are absolutely separated from each other.

2.3.2.2 Advantages

a) Common Advantages of the ARCP Circuits:

The ARCP circuits introduced in this section are attributed with the advantage that the auxiliary switches need only withstand half of the DC bus voltage and are turned on and off under zero current conditions. Hence, the losses of the auxiliary circuit are only conduction loss and switch capacitive turn-on loss. Since the blocking voltage of the auxiliary switches is now half of DC bus voltage, the capacitive turn-on loss of the auxiliary circuit is also reduced compared to the circuit introduced in 2.3.1. The auxiliary circuit is able to operate with any SVM technique. For commutation of phase currents only the related auxiliary branches are activated resulting in minimizing the auxiliary circuit losses. However, six auxiliary switches are required. This is usually unaffordable except in very high power applications.

b) Additional Advantage of the ACS-HW-ARCP Circuit:

Beside the advantages of the conventional ARCP circuit, the ACS-HW-ARCP circuit proposed in this section has also this advantage that switching control of the circuit for simultaneous commutation of two phase currents is easier. The auxiliary branches of this circuit are not completely separated. The auxiliary inductor is a common component for the three phases. Advantage of this structure appears in all three stages of simultaneous commutation of the phase currents with different magnitude as shown in Fig. 2.13. At the end of the charging state, when the switches S_{bn} and S_{cn} are turned off, I_{boost} is divided equally between phases b and c, which is not possible by the conventional circuit. Equal division of the boost current between phases b and c leads to the same voltage variation at these phases, so that the switches S_{bp} and S_{cp} can be turned on simultaneously under ZVS conditions. In the discharging state, currents I_{bx} and I_{cx} reach zero at the same time, because the node voltage V_b and V_c are slightly different. So turn-on and turn-off times in both main and auxiliary circuits are synchronised. This reduces the complexity of the switching control timing in the proposed ARCP to some extent.

2.3.2.3 Another Topology of ACS-HW-ARCP Concept

The center point of the DC side can be realized either by a double capacitor as the case of conventional ARCP circuit or by two coupled inductors at each phase having the same inductances as shown in Fig. 2.14(a). In this circuit the auxiliary inductor currents are divided equally between the coupled inductors, so the current of the auxiliary devices is halved causing their conduction loss to be halved too.

This circuit can also be simplified for the rectifying applications, where ZVS turn-on synchronization of switches using the switching sequence $S_{1(pnn)} - ACA - S_{7(ppp)} - S_{2(ppn)}$ is desired, as shown in Fig. 2.14(b).

2.3.2.4 Problems and Solutions

The pole commutation concept seems to solve the problems of the ACS-HW-ARC circuit described in 2.3.1.3. The auxiliary circuit is also able to accelerate

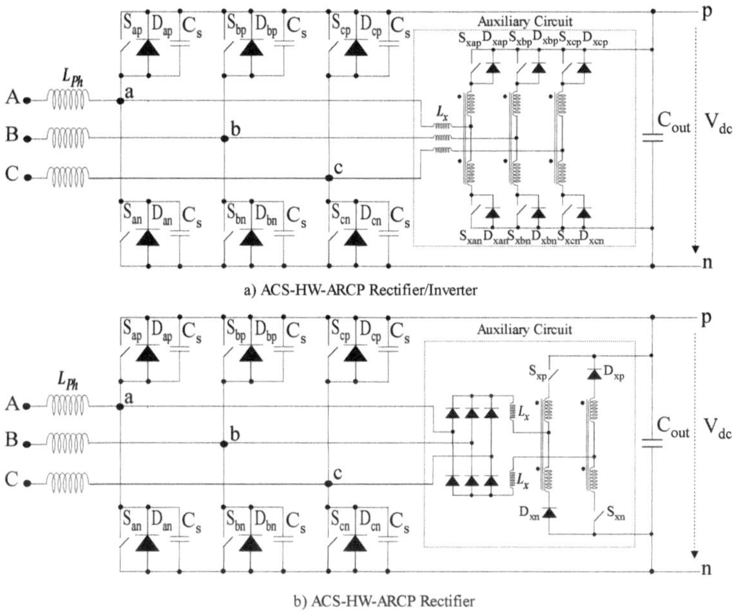

Fig. 2.14: Different realization of the ACS-HW-ARCP concept.

the switch-to-diode commutations at small phase currents. The problem which still exists and should be discussed is the necessity of the circuit to the boost energy. This energy should compensate the losses of the auxiliary circuit and the deviation of the DC side center point voltage from zero as well. Another important point which should be also noted in the half-wave concepts is that in the resonant state the node voltages should *resonate to* a level a quite bit higher than the opposite DC rail voltage to provide the complete ZVS conditions for the on-going switches by conduction of their antiparallel diode. In other words for conduction of the antiparallel diode of the on-going switches some surplus current is necessary at the end of resonant state. Since in a half-wave resonant network the resonant current theoretically decays to zero after a half resonance period, enough boost energy is required at the beginning of the resonant state for providing the necessary surplus current at the end of this state.

Consider the resonant state of the example of Fig.2.13(a). To provide the complete ZVS conditions for the switches S_{ap} and S_{cp}, their antiparallel diodes D_{ap} and D_{cp} have to start conducting and should remain in the conducting state for a certain duration until the switches are turned on. For this purpose the node voltages V_b and V_c should resonate to a level beyond the positive DC rail voltage. The auxiliary inductor also has to charge to a value higher than the sum of phase currents to provide the necessary initial resonant energy. This results in additional conduction loss in the auxiliary devices and extra turn-off loss in the main switches turned off under the surplus currents at the end of boost charging state.

This additional switching action is a serious drawback of the HW-topologies. In addition, switching control timing in the ZVS commutation is also very critical. As a result, more turn-off losses and control complexity are introduced, and their efficiency improvement is quite limited.

The above discussion suggests that it would be advantageous to introduce an auxiliary circuit which has no need for the boost energy and additional switching actions and can ensure the ZVS conditions at any load condition.

The purposes of additional main switching actions in the previous HW-topologies were:

1. To charge/discharge the auxiliary inductors at zero voltage vectors where no voltage is available at the auxiliary circuit in the ACS-HW-ARC circuit,

2. To charge the auxiliary inductor current much higher than is required for diodes turn-off in both circuits.

By introducing additional commutation components at the auxiliary circuit for recovering the commutation energy and by connecting the auxiliary inductors to the DC rails during the resonant state the circuit can be improved, so that both problems are solved altogether. Connection of the auxiliary inductors to the DC rails introduces a new family of auxiliary circuits. In this concept the ZVS conditions are achieved after a Quarter-Wave (QW) oscillation of the resonant circuit. For this reason the circuits utilizing this concept are named Quarter-Wave circuits compared to Half-Wave circuits. These circuits are the second group of ASSC converters discussed in this work.

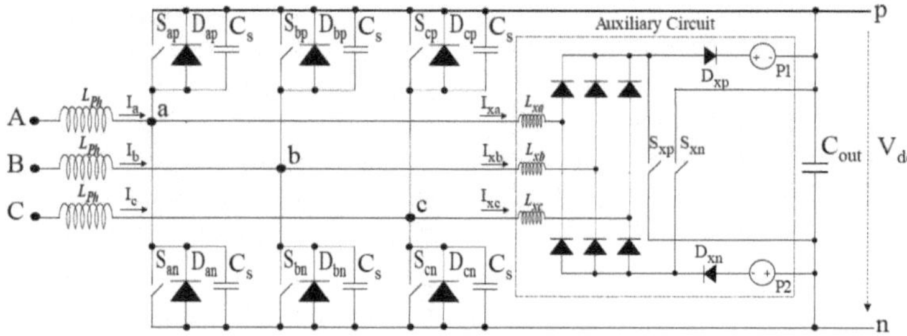

Fig. 2.15: The quarter-wave auxiliary resonant commutated converter.

2.3.3 AC-Side Quarter-Wave Auxiliary Resonant Commutated Converter

One of the topologies utilizing the quarter-wave concept is shown in Fig. 2.15 and named the AC-Side Quarter-Wave Auxiliary Resonant Commutated (ACS-QW-ARC) Circuit.

In this topology, the six main switches are divided into two groups: three upper switches and three bottom switches. S_{xp} is used to provide ZVS conditions for the upper switches S_{ap}, S_{bp} and S_{cp}, while S_{xn} is used to provide these conditions for the bottom switches S_{an}, S_{bn} and S_{cn}. Like the ACS-HW-ARC circuit, the auxiliary circuit is also a three-phase boost rectifier operating in discontinuous mode. The auxiliary switches S_{xp} and S_{xn} should be turned on only for a short time. In this topology the ZVT conditions can be achieved under any circumstances for all main switches. To minimize the number of switching and losses in the auxiliary circuit, a modified space vector scheme (see section 2.3.3.2) is used to synchronize the turn-on instants of main switches where it is possible.

2.3.3.1 Auxiliary Circuit Operation

In the following discussion, it is assumed that the converter is operating in the first half of the sector I of voltage vector hexagon. According to the modified SVM described in the next section, during the first half of the sector I, S_{ap} is on, S_{an} is off, and the remaining four switches are modulated. Within this interval, the converter operates in the rectifier mode if the currents I_a, I_b and I_c

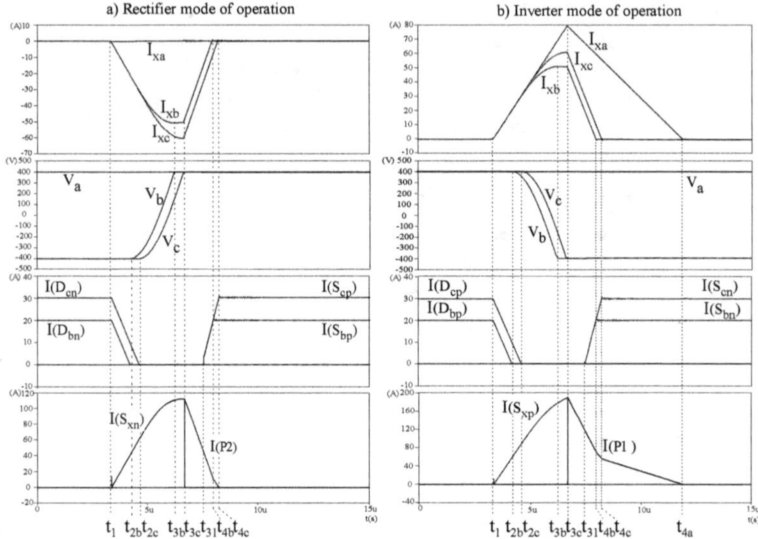

Fig. 2.16: ACS-QW-ARC converter waveforms during an ACA.

are positive, negative and negative and operates in the inverter mode if they are negative, positive and positive respectively. Fig. 2.16 shows the converter waveforms during an ACA in both operation modes.

a - Operation as Rectifier:

The same transition $S_{1(pnn)} \rightarrow S_{7(ppp)}$ as used in the example of the ARC-HW-ARCP circuit is considered. Phase currents I_a, I_b and I_c flowing through the diodes D_{ap}, D_{bn} and D_{cn} are positive, positive and negative, respectively. Since there is no need for the boost energy in this circuit, the switches S_{ap}, S_{bn} and S_{cn} can be on or off depending on the application. The auxiliary switches are off, and there is no current in the auxiliary circuit.

t_1 to t_{2b}/t_{2c}:

The ACA starts with turning on the auxiliary switch S_{xn} at t_1. The diode D_{xn} is blocked, and full DC voltage is applied across the auxiliary inductors L_{xb} and L_{xc}. The auxiliary currents I_{xb} and I_{xc} increase in the negative direction linearly. The current of the diodes D_{bn} and D_{cn} also decreases linearly. Since no voltage is applied across L_{xa}, I_{xa} remains zero. At t_{2b}, I_{xb} reaches the phase current I_b and D_{bn} turns off softly. The same happens to D_{cn} at t_{2c}. If the switches S_{bn} and S_{cn} are closed due to modulation requirements as mentioned before, they can be turned off under the zero current and voltage conditions any time before their antiparallel diode D_{bn} and D_{cn} turns off.

t_{2b}/t_{2c} to t_{3b}/t_{3c}:

After turn-off of the diodes D_{bn} and D_{cn} at t_{2b} and t_{2c} respectively, L_{xb} at phase b and L_{xc} at phase c immediately start to resonate with the node capacitances. After a quarter of resonance period the current of the resonant inductors I_{xb} and I_{xc} becomes maximum and the node voltages V_b and V_c reach the positive DC rail voltage. The energy stored in the auxiliary inductor is now used to further increase the node voltages beyond the positive DC rail voltage until D_{bp} and D_{cp} turn on and ZVS conditions for turning on the switches S_{bp} and S_{cp} are achieved at t_{3b} and t_{3c}. Due to enough surplus current in the auxiliary inductors , D_{bp} and D_{cp} continue conducting for a sufficient duration ensuring the ZVS conditions for S_{bp} and S_{cp} at any load condition. This is why the auxiliary circuits employing QW-concept have no need for the boost energy. At t_{3c}, S_{bp} and S_{cp} are turned on and S_{xn} is turned off simultaneously.

t_{3c} to t_{4b}/t_{4c}:

After S_{xn} is turned off, the auxiliary currents charge the parasitic capacitance of the switch and the diode D_{xn} begins conducting. This reverses the voltage across the auxiliary inductors and transfers the circuit to the discharging state resulting in decreasing the auxiliary currents I_{xb} and I_{xc}. The DC supply $P1$ also helps to recover the energy of the auxiliary inductors, but basically, it is only necessary for the inverter mode of operation as it will be described in the following discussion. At t_{31}, when the auxiliary currents become smaller than the phase currents, S_{bp} and S_{cp} begin conducting. At t_{4b} and t_{4c} the auxiliary

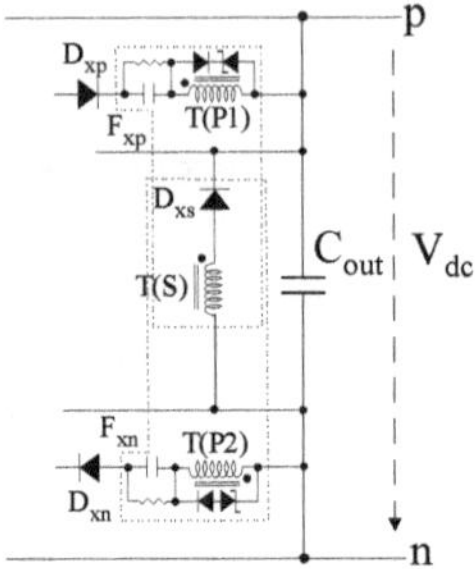

Fig. 2.17: Energy feedback circuit for inverter mode of operation.

currents I_{xb} and I_{xc} become zero, D_{xn} blocks and the ACA ends.

b - Operation as Inverter:

When the converter operates in the inverter mode, the sequence of switching changes (see the next section). Because the AC currents I_a, I_b and I_c are now reversed, the auxiliary circuit has to be activated for the transition $S_{7(ppp)} \rightarrow S_{1(pnn)}$ when the upper diodes D_{bp} and D_{cp} turn off, and the bottom switches S_{bn} and S_{cn} should be turned on under ZVS conditions.

At t_1 the auxiliary switch S_{xp} is turned on. The DC voltage is now applied across all three auxiliary inductors causing L_{xa} to be also charged useless. This problem, which is a serious drawback for using this converter as an inverter, cannot be avoided in this circuit.

The resulting inverter operation in charging and resonant states is similar to the rectifier operation except that I_{xa} also increases for nothing. At t_{3c}, S_{bn} and S_{cn} are turned on under ZVS conditions and S_{xp} is also turned off. The auxiliary currents are delivered to D_{xp}. This reverses the voltage across the auxiliary inductors L_{xb} and L_{xc}, so the discharging state begins as usual for these inductors. But the voltage across L_{xa} falls down to zero. The DC supply is now necessary to recover the energy stored in L_{xa}. This can be either an auxiliary power supply for control and drive circuits or an energy feedback circuit F_{xp} as shown in Fig. 2.17.

2.3.3.2 Modified SVM Scheme

As mentioned, in order to minimize the number of switching actions in the auxiliary circuit, a modified space vector scheme is used for synchronizing the ZVS turn-on of the main switches when it is possible. This scheme changes the switching sequence within each 60° interval according to the relative magnitude of the phase currents. The switches connected to the phase with the largest current magnitude are left on/off, and the others work in PWM operation. So each switch is assigned two 60° no switching intervals within the fundamental period.

Assume that reference voltage vector $\vec{V}_{ref}$ is located in sector I as shown in Fig. 2.7. The modified SVM is described as follows:

a - Rectifier mode of operation:

In this mode, the phase current directions are all similar to the node polarities in one of the switching states $S_{1(pnn)}$ and $S_{2(ppn)}$. Since the switching state with the node voltage polarities *similar to* the phase current directions has *the largest number of conducting diodes*, actuation of the auxiliary circuit is placed at the end of duty cycle of this switching state. With this strategy, the required diode-to-switch current commutations can be easily synchronized, thereby the number of switching actions in the auxiliary circuit is also minimized. Note that this is a general strategy: *where synchronization of diode-to-switch current commutations is concerned, ACA should be placed at the end of the duty cycle of the switching state with the largest number of conducting diodes.*

ACA then transfers the circuit to the switching state with the largest number of conducting switches. Since from a switching state with the largest number of conducting switches other desired switching states can be easily attained by opening the switches, this strategy minimizes the number of required ACAs. Fig. 2.18(a) shows the modulation scheme in the rectifier mode. Consider the case $I_a > 0$, $I_b < 0$ and $I_c < 0$ for example: since direction of the phase current with the largest magnitude is always opposite to the direction of others, I_a with the direction (positive) opposite to the direction of I_b and I_c (negative) has the largest magnitude, so switching at this phase should be avoided. The phase current directions, i.e. positive, negative and negative (pnn), correspond to the node voltage polarities in the switching state $S_{1(pnn)}$. Since $S_{1(pnn)}$ have the

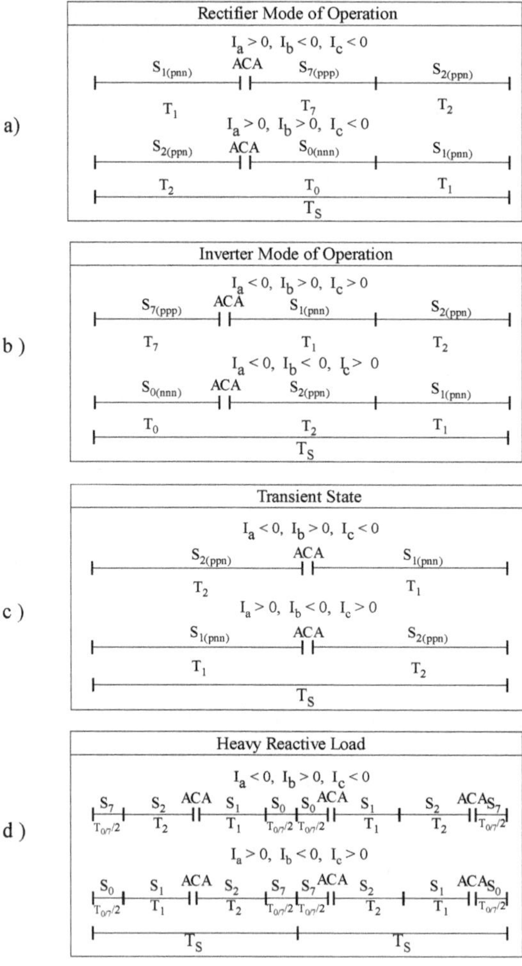

Fig. 2.18: The modified space vector modulation ($\vec{V}_{ref}$ in sector I).

largest number of conducting diodes (D_{ap}, D_{bn} and D_{cn}), ACA is placed at the end of duty cycle of this switching state. After ACA, the circuit is transferred to $S_{7(ppp)}$ with the largest number of conducting switches (S_{bp} and S_{cp}). Other transitions $S_{7(ppp)} \rightarrow S_{2(ppn)} \rightarrow S_{1(pnn)}$ are self-commutations[12] which can be accomplished by opening of the switches S_{cp} and S_{bp} one at a time. The switch S_{ap} remains closed during the entire switching cycle and the current I_a flows through the diode D_{ap}.

b - Inverter mode of operation:

In this mode of operation, the phase current directions are all opposite to the node polarities in one of the switching states $S_{1(pnn)}$ and $S_{2(ppn)}$ when $\vec{V}_{ref}$ is located in voltage sector I. In this case the switching state with the node voltage polarities *opposite to* the phase current directions has *the largest number of conducting switches*, and one of the zero switching states has*the largest number of conducting diodes*. Synchronization of diode-to-switch current commutations is now achieved if ACA is placed at the end of duty cycle of zero switching state. The sequence is then followed as shown in Fig. 2.18(b).

c - Transient state or heavy reactive load:

In these conditions, the phase current directions are neither all similar nor all opposite to the node polarities in one of the switching states $S_{1(pnn)}$ and $S_{2(ppn)}$. In transient states, as it will be described in chapter 6, zero voltage vector is eliminated in order to drive the phase currents to new current references as fast as possible. Therefore current commutation is only necessary at one phase and can be performed with only one ACA (see Fig. 2.18(c)). In heavy reactive loads when the current phase angle φ is larger than 30°, the SVM scheme is switched back to the standard voltage space vector modulation [61] for the region $\left(\phi < (\varphi - 30^o)\right)$. In this region turn-on synchronization is not possible and the auxiliary circuit averagely has to be activated more than once during each sampling cycle (see Fig. 2.18(d)).

If the switching sequence starts with the auxiliary circuit action (ACA), a constant frequency clock signal can be used for switching of the auxiliary switches

[12] Assuming the magnitude of phase currents large enough.

in the rectifier and inverter mode of operation as well as in the transient state. In this way, control timing of the auxiliary circuit will be simplified.

2.3.3.3 Problems and Solutions

Use of the QW-concept in simultaneous commutation of two phase currents has this advantage that both current commutations are accomplished with only one switching action in the auxiliary circuit, there is no need for the boost energy and the commutation processes (charging, resonant and discharging stages) are faster, but in return the turn-off conditions at the auxiliary circuit are very hard. Although the ACS-QW-ARC circuit shows advantages in solving the problems related to boost energy, the circuit has two basic problems:

First, the auxiliary switches should cut off a current as large as the sum of the auxiliary current peaks which is certainly a very hard turn-off action. This excludes use of IGBTs for the auxiliary switches. In case of Power MOSFETs, the relatively large capacitive turn-on loss of MOSFETs decreases the attractiveness of the circuit for high power applications.

The second problem of this circuit is the inverter mode of operation which is not convenient. Serious drawbacks of the inverting operation of this circuit are:

1. Useless charging of the auxiliary inductor connected to the phase with the largest current magnitude results in further losses and higher current peak in the auxiliary switches,

2. The use of a DC power supply or an energy feedback circuit for recovering the energy of the auxiliary inductor mentioned in 1 increases the complexity of the auxiliary circuit.

The problems related to inverter mode of operation are solved if the auxiliary branch connected to the phase with the largest current magnitude is not activated during the commutation process. This suggests the independent commutation of phase current, i.e. pole commutating concept, as was the case of ASC-HW-ARC circuit. Use of this concept in the ACS-QW-ARC circuit leads to a new topology which is proposed in the next section.

However, it may be desirable to use the converter only in the rectifier mode of operation. In this case the auxiliary circuit can be simplified to the topology

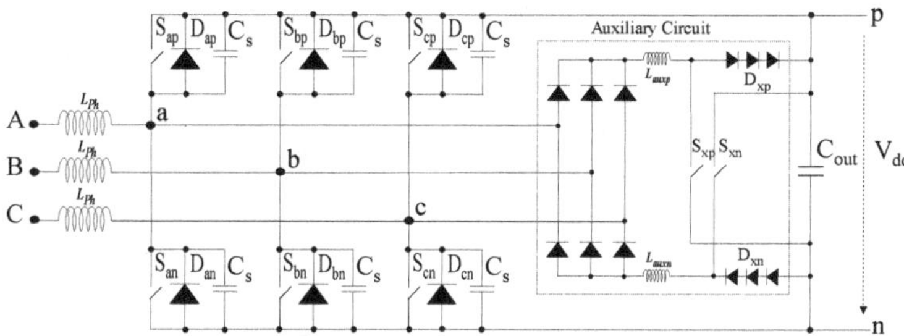

Fig. 2.19: The AC-side quarter-wave auxiliary resonant commutated rectifier

shown in Fig. 2.19. The operation of this circuit is similar to the rectifier operation described in section 2.3.3.1.

2.3.4 AC-Side Quarter-Wave Auxiliary Resonant Commutated Pole Converter

Like the ACS-HW-ARCP topology, six auxiliary switches are used to provide the independent pole current commutation for every phase. The QW-concept is realized in this topology with connecting the auxiliary inductors to the DC rails. Fig. 2.20 shows the AC-Side Quarter-Wave Auxiliary Resonant Commutated Pole (ACS-QW-ARCP) circuit considering all components ideal[13)] .

Similar to ACS-HW-ARCP converter, this converter can also be controlled with any SVM scheme. The auxiliary circuit operates like the ACS-QW-ARC circuit at each phase, providing complete zero voltage switching conditions for the on-going switches without overcharging the auxiliary inductors.

2.3.4.1 Auxiliary Circuit Operation in Inverter Mode

Fig. 2.21 shows waveforms of the converter operation in the inverter mode during an auxiliary circuit action. The typical turn-on commutation in inverter mode is the commutation of I_b and I_c from D_{bp} and D_{cp} to S_{bn} and S_{cn}, which transfers the circuit from the switching state $S_{7(ppp)}$ to $S_{1(pnn)}$ in the switching

[13)] In case of real components the circuit becomes to some extent more complex as shown in section 2.3.4.2.

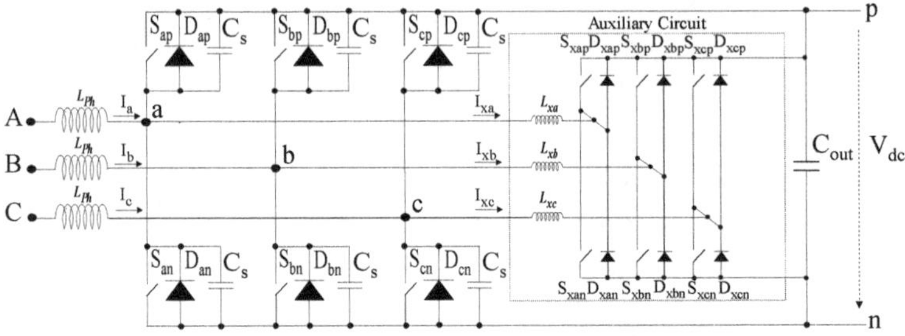

Fig. 2.20: The quarter-wave auxiliary resonant commutated pole converter

sequence $S_{7(ppp)} - ACA_{b,c} - S_{1(pnn)} - S_{2(ppn)}$, while $I_a < 0 < I_b < I_c$.

The ACA at phases b and c starts with the turn-on of S_{xbn} and S_{xcn} at t_1. The auxiliary inductors L_{xb} and L_{xc} charge under full DC bus voltage until the currents through D_{bp} and D_{cp} decrease to zero at t_2 and t_3. The charging state takes half of the time required for the half-wave resonant circuit, since full DC bus voltage is applied across the auxiliary inductors. After the diodes turn off, the resonant state immediately starts and the node voltages V_b and V_c resonate to the negative DC rail. S_{bp} and S_{cp} can be turned on under ZVS conditions when the node voltages, V_b and V_c, reach the negative DC rail voltage at t_4 and t_5. The auxiliary switches S_{xbn} and S_{xcn} are also turned off at t_4 and t_5 respectively. The auxiliary currents, I_{xb} and I_{xc}, are then delivered to D_{xbp} and D_{xcp} causing the voltage across the auxiliary inductors to be reversed and I_{xb} and I_{xc} to decrease linearly. The ACA ends at each phase when the corresponding auxiliary current reaches zero (t_6 and t_7).

Other transitions, $S_{1(pnn)} \rightarrow S_{2(ppn)} \rightarrow S_{7(ppp)}$ take place by opening the switches S_{bn} and S_{cn}, one at a time. Note that in the above example, the soft commutation of the phase currents I_b and I_c take place independently. Therefore, one would use the switching sequence $S_{7(ppp)} - ACA_c - S_{2(ppn)} - ACA_b - S_{1(pnn)}$ for example. As mentioned in section 2.3.2.1 the resulting losses of the auxiliary

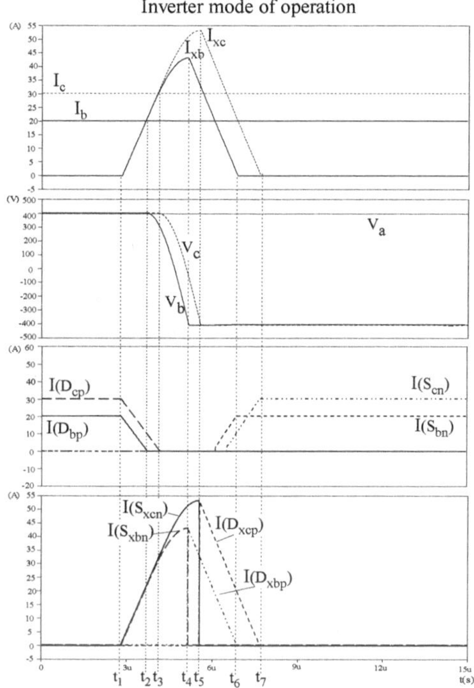

Fig. 2.21: Waveforms of ACS-QW-ARCP during an ACA.

circuit remain the same, but the switching sequence with the separate commutation of phase currents is advantageous, when it is used in a switching pattern like 127721 (see also chapter 4).

2.3.4.2 Design Considerations

After an ACA, when the auxiliary diodes turn off, the auxiliary circuit should be functionally disconnected from the main power stage and the converter must resume its PWM operation until the next ACA. To achieve these conditions, the auxiliary current paths providing ZVS conditions for the upper and bottom switches should be separated from each other as shown in Fig. 2.22(b). This is because of the following reason: suppose in the circuit of Fig. 2.22(a) at the end of an ACA at phase c the auxiliary diode D_{xcp} turns off and the auxiliary circuit should be disconnected from the main circuit automatically. When D_{xcp} turns off, the parasitic capacitance of the auxiliary switch S_{xcn}, which is precharged to V_{dc}, starts to resonate with the auxiliary inductor at phase c. The voltage across S_{xcn} is reduced to zero and D_{xcn} begins conducting. Conduction of this diode provides a freewheeling path for the auxiliary current I_{xc} through the closed switch S_{cn}. Since the resistance of this freewheeling path is very small, the auxiliary current cannot decrease quickly and flows through S_{cn} until the switch is turned off. Therefore, the auxiliary current paths related to the upper and bottom switches are separated from each other as shown in Fig. 2.22(b) and a diode in series with the auxiliary inductors is used to block the freewheeling current path.

Furthermore a series connection of more than one diode should be designed for the auxiliary diodes to ensure the disconnection of the auxiliary circuit from the main circuit in normal PWM operations. For example suppose that D_{cp} is conducting the phase current I_c. Conduction of the diode causes the node voltage V_c to rise by some volts higher than the positive DC rail voltage. If the increase of V_c beyond the DC rail voltage becomes larger than the threshold voltage of D_{xcp}, the latter diode would also conduct a part of the phase current as shown in Fig. 2.22(b).

2.3.4.3 Problems and Solutions

The QW concept of ARCP circuit needs no extra main switch turn-off action, its auxiliary currents are automatically limited to the phase currents at the end of charging state. The auxiliary and main switches are turned off simultaneously resulting in a reduction in complexity of control timing of the circuit. The necessary time for an ACA is also almost half of the time required for a half-wave concept.

The aforementioned advantages are at the expense of hard turn-off switching conditions for the auxiliary switches. Since the auxiliary switches are usually implemented with Power MOSFETs because they should be switched much faster than the main switches, the turn-off loss may be ignored, but their conduction and capacitive turn-on losses are to be considered. This problem limits the application of this circuit to medium power range. The solution would be improvement of the switching conditions at the auxiliary circuit which is left to future works.

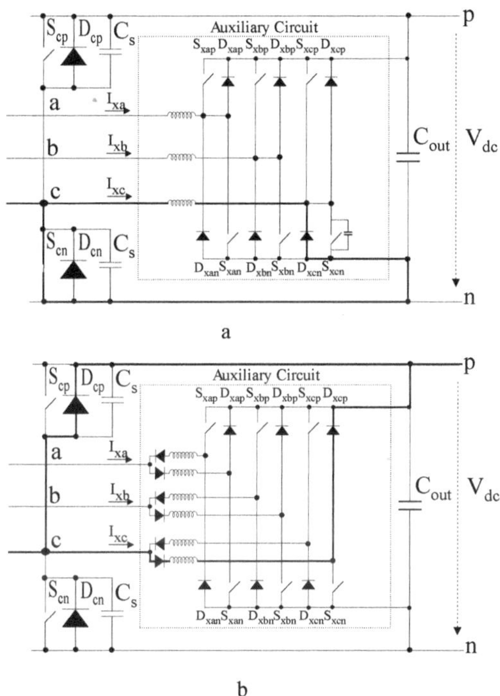

Fig. 2.22: Design consideration of the ACS-QW-ARCP converter.

2.4 Summary of Chapter

This chapter discusses first the hard switching and soft switching (ZVS and ZCS) techniques. The results obtained from simulations using physic-based dynamic models of IGBT and Power diode show that the ZVS technique reduces the total switching losses including the diode reverse-recovery losses more effectively and eliminates the problems related to the reverse-recovery-effect of power diodes as well.

The discussion of this chapter continues with the introduction of AC-Side Soft Commutated (ASSC) PWM Three-Phase converters. The ASSC topologies are categorized in this chapter in two ways:

1. *Half-Wave (HW) and Quarter-Wave (QW) concepts:* the auxiliary resonant network of HW-topologies provides the ZVS conditions after a half resonance cycle, while the resonant network of QW-topologies provides these conditions after a quarter resonance cycle (see Fig. 2.23).

2. *Commutated Pole and Non-Pole Commutated concepts:* in commutated pole topologies (Fig. 2.24(b and d)) the soft current commutation occurs independently in every phase. In contrary, in non-pole commutated topologies (Fig. 2.24(a and c)) the soft commutations of phase currents are not independent from each other. Hence, the converter should be controlled by a certain PWM control algorithm which is developed to meet the requirement of auxiliary circuit operation.

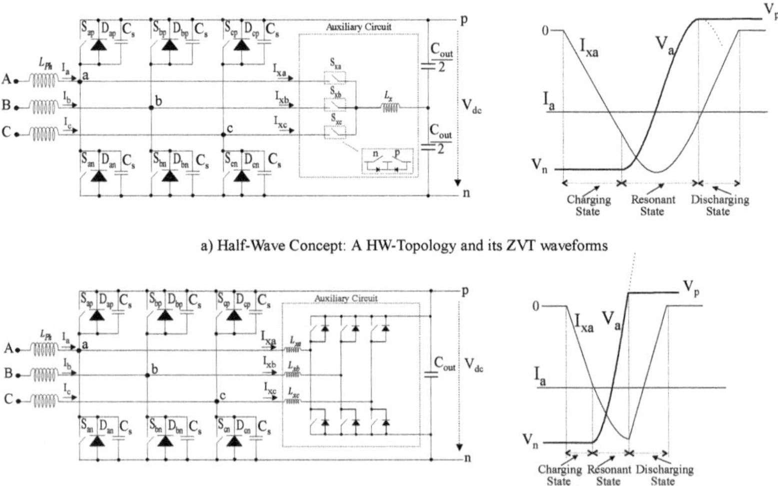

a) Half-Wave Concept: A HW-Topology and its ZVT waveforms

b) Quarter-Wave Concept: A QW-Topology and its ZVT waveforms

Fig. 2.23: Half-Wave and Quarter-Wave concepts.

Starting with the AC-Side Half-Wave Auxiliary Resonant Commutated (ACS-HW-ARC) Converter (Fig. 2.24(a)) which has only one semiconductor switch at its auxiliary circuit, the discussion enumerates the disadvantages of this converter respecting lack of input voltage at the auxiliary circuit, when switching one of the zero voltage vectors, $\vec{V}_{7(ppp)}$ or $\vec{V}_{0(nnn)}$, and the problems relating to synchronous diode-to-switch current commutation at all three phases. For solving the problems of ACS-HW-ARC circuit, its commutated pole concept, i.e. the AC-Side Half-Wave Auxiliary Resonant Commutated Pole (ACS-HW-ARCP) Converter (Fig. 2.24(b)) is suggested. Although this circuit can be considered as the best candidate for high power applications, it has still two problems: 1-the necessity of the circuit to high boost energy for creating complete ZVS conditions, which results in the complexity of control timing of the converter, and 2-introduction of three four-quadrant switches in the auxiliary circuit which practically need six IGBTs. Using six active auxiliary switches, the commutated pole circuits are able to commutate the phase currents independent from each other. This ability is very advantageous because it allows the modulation technique to be optimized from the losses point of view. It will be shown in chapter 4 that it is worthy to have an optimized SVM technique at expense of six auxiliary switches.

The problems regarding boost energy can be solved by employing Quarter Wave-concept (Fig. 2.24(c and d)). The Quarter Wave-concept ensures complete ZVS conditions at any load condition and its voltage transition is twice as fast as in Half Wave-concept. These advantages, however, are achieved at the expense of losing ZCS turn-off conditions and much larger current peak for the auxiliary switches. This limits the application of Quarter Wave-circuits to low and medium power ranges.

Table 2.2 presents a qualitative comparison among the discussed converters. The properties of the circuits are valued with plus (+) and minus (-) points. To show the comparison also quantitative, some of the properties are valued with more than one plus or minus point.

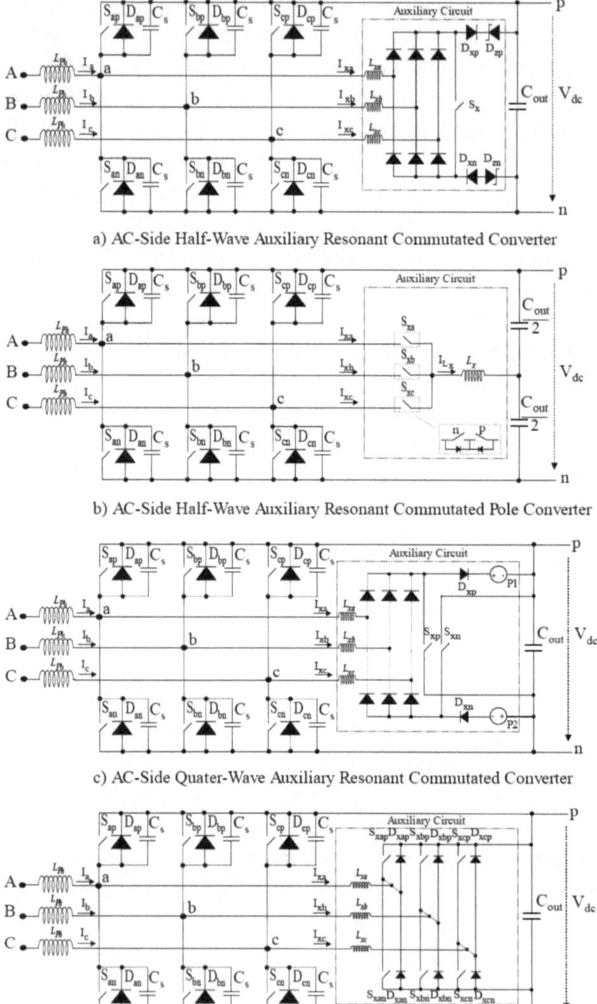

a) AC-Side Half-Wave Auxiliary Resonant Commutated Converter

b) AC-Side Half-Wave Auxiliary Resonant Commutated Pole Converter

c) AC-Side Quater-Wave Auxiliary Resonant Commutated Converter

d) AC-Side Quarter-Wave Auxiliary Resonant Commutated Pole Converter

Fig. 2.24: Three-Phase ASSC PWM Converters.

Qualitative Comparison of ASSC PWM Circuits				
Property	ACS-HW-ARC	ACS-HW-ARCP	ACS-QW-ARC	ACS-QW-ARCP
Disturbance in SVM	yes (-)	no (+)	no (+)	no (+)
Limitation on SVM Technique	yes (-)	no (+)	yes (-)	no (+)
Reduction of DC Bus Utilization*	yes (-)	no (+)	no (+)	no (+)
Number of Auxiliary Switches	1 switch (+,+)	6 switches (-)	2 switches (+)	6 switches (-)
Rectifying and Inverting Operations	similar (+)	similar (+)	different (-)	similar (+)
Switching Conditions at Auxiliary Circuit	ZCS turn-off (+) capac. turn-on(-,-)	ZCS turn-off (+) capac. turn-on(-)	hard turn-off (-,-) capac. turn-on(-,-)	hard turn-off (-) capac. turn-on(-,-)
Additional Switching Actions in Main Circuit	for undesired state (-) for boost energy (-)	for boost energy (-)	no add. actions (+)	no add. actions (+)
Auxiliary Switch Implementation	with IGBT (+)	with IGBT (+)	with MOSFET (-)	with MOSFET (-)
Assessment	-2	+3	-3	0

*Caused by the modifications performed in SVM

Table 2.2: Qualitative assessment of the ASSC PWM circuits.

3. Analysis and Comparative Study of the AC-Side Soft Commutated PWM Converters

In chapter 2, the AC-Side Soft Commutated (ASSC) PWM converters are introduced and their features are also discussed. In this chapter these circuits will be studied more precisely. The reason for separate discussion of the circuits in two chapters has to do with the kind of discussions. In chapter 2, it was descriptively pointed out what the benefits of the ASSC-PWM converters are and how they operate. However, in this chapter the circuits will be mathematically analysed and the expressions of nodes voltage and of auxiliary currents during the transition stages (charging, resonant and discharging states) will be explored for all topologies under study. The expressions describing device stresses and basic design equations will be derived as well. The circuits will also be evaluated from the loss point of view in an example considering two widely used power semiconductors, the IGBT and Power MOSFET.

3.1 Circuit Analysis

In analysing the circuits in this section, the components are assumed to be ideal, except where it is mentioned. Furthermore the same operation conditions as considered for discussion of the rectifier mode operation of converters in the last chapter, i. e. the reference voltage vector $\vec{V}_{ref}$ located inside the first half of sector I (see Fig. 2.7) and $I_a > 0 > I_b > I_c$, are assumed here. The phase currents are also assumed to remain constant during the switching cycle and consider the directions shown on the figures of the circuits positive. The center point of the DC-side is the reference point for the node voltages V_a, V_b and V_c. The analyses are applicable to the duration of an auxiliary circuit action (ACA).

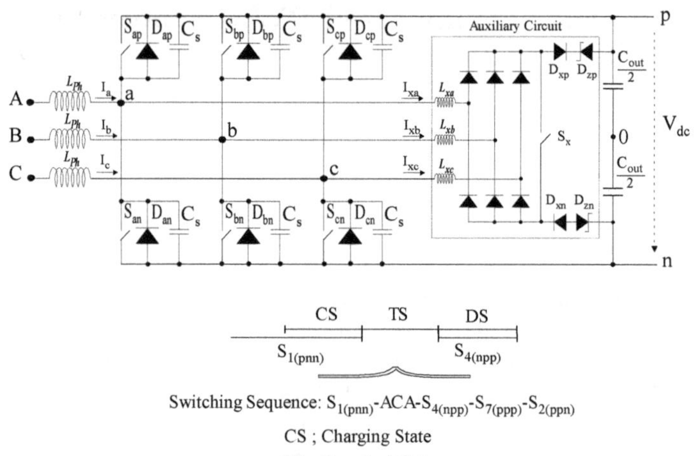

Fig. 3.1: ASC-HW-ARC PWM converter and the switching sequence.

The characteristics of the resonant circuit used in this section are defined as follows:

$$\left\{\begin{array}{lcl} L_x & = & L_{xa} = L_{xb} = L_{xc}, \\ \omega_r & = & \dfrac{1}{\sqrt{2\,C_s\,L_x}}, \\ Z_r & = & \sqrt{\dfrac{L_x}{2\,C_s}}. \end{array}\right. \tag{3.1}$$

3.1.1 The ACS-HW-ARC Circuit

The ACS-HW-ARC circuit is shown in Fig. 3.1 along with the switching sequence used. The analysis is carried out for the transition between the switching states $S_{1(pnn)}$ and $S_{4(npp)}$. In this transition the phase currents I_a, I_b and I_c are commutated from D_{ap}, D_{bn} and D_{cn} to S_{an}, S_{bp} and S_{cp} respectively by activating

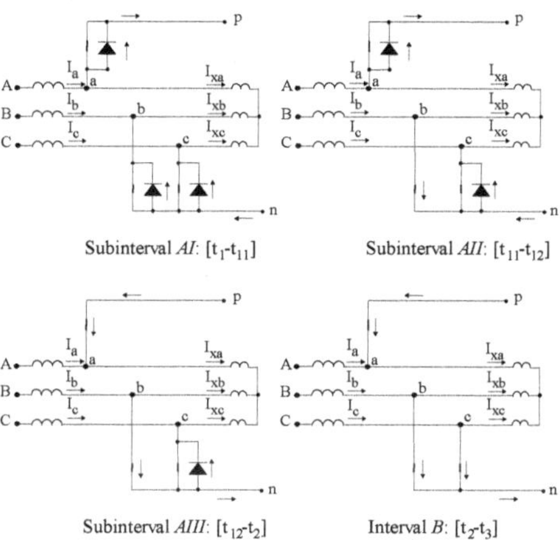

Fig. 3.2: Topological sequence of the charging and boost charging states (ACS-HW-ARC circuit)

the auxiliary circuit. For the simulative waveforms of the node voltages and of the auxiliary currents see Fig. 3.5 (see also Fig. 2.9(a)).

3.1.1.1 Analysis

a) Charging and Boost Charging States:

The charging state starts at $t_1 = 0$ when the auxiliary switch S_x is turned on. The node voltages at this instant are $V_a(t_1) = V_{dc}/2$, $V_b(t_1) = -V_{dc}/2$ and $V_c(t_1) = -V_{dc}/2$. With turning on the auxiliary switch, the auxiliary currents I_{xa}, I_{xb} and I_{xc} start to increase linearly according to the following expressions:

$$I_{xa}(t) = \frac{2}{3}\frac{V_{dc}\,t}{L_x}, \tag{3.2}$$

$$I_{xc}(t) = I_{xb}(t) = -\frac{1}{3}\frac{V_{dc}\,t}{L_x}. \tag{3.3}$$

This state consists of three subintervals as shown in Fig. 3.2. During the subinterval AI, $[t_1 - t_{11}]$, I_{xb} reaches the phase current I_b leading to turn-off of the diode

D_{bn}. The auxiliary inductor L_{xb} continues charging (boost energy). Similarly this proceeds for phases a and c during the subintervals AII and $AIII$ ($[t_{11} - t_{12}]$ and $[t_{12} - t_2]$). The times t_{11}, t_{12} and t_2 are determined by:

$$t_{11} = 3\,\frac{L_x\,|I_b|}{V_{dc}}, \tag{3.4}$$

$$t_{12} = \frac{3}{2}\,\frac{L_x\,|I_a|}{V_{dc}}, \tag{3.5}$$

$$t_2 = 3\,\frac{L_x\,|I_c|}{V_{dc}}. \tag{3.6}$$

During the interval B, $[t_2 - t_3]$ (boost charging state), the auxiliary current I_{xa} attains the value $(I_a + I_{boost})$. The switches S_{ap}, S_{bn} and S_{cn} conducting the surplus currents $(I_{xa} - I_a)$, $(|I_{xb} - I_b|)$ and $(|I_{xc} - I_c|)$ are turned off simultaneously at this time and the circuit is transferred to the resonant state. The time t_3 and the auxiliary currents I_{bx} and I_{xc} at t_3 are given by:

$$t_3 = \frac{3}{2}\,\frac{L_x\,(I_a + I_{boost})}{V_{dc}}, \tag{3.7}$$

$$I_{xb}(t_3) = I_{xc}(t_3) = -\frac{1}{2}\,(I_a + I_{boost}). \tag{3.8}$$

The times t_2 and t_3 represent the end of charging and boost charging states in Fig. 2.9(a). Note, that $I_a + I_{boost}$ should be larger than $2\,|I_c|$ ($|I_c| > |I_b|$). The whole charging duration Δt_{ch} is calculated as follows:

$$\Delta t_{ch} = t_3 - t_1 = \frac{3}{2}\,\frac{L_x\,(I_a + I_{boost})}{V_{dc}}. \tag{3.9}$$

b) Resonant State:

This stage comprises three subintervals. The topological sequence of the converter during the resonant state is shown in Fig. 3.3. During the subinterval CI, $[t_3 - t_{31}]$, all of the capacitances at nodes a, b and c take effect and node voltages V_a, V_b and V_c resonate to the opposite DC rail voltage. The node voltages can be described during this subinterval as follows:

$$\begin{cases} V_a(t-t_3) - V_b(t-t_3) = L_x\left(\frac{d}{dt}I_{xa}(t-t_3) - \frac{d}{dt}I_{xb}(t-t_3)\right), \\ V_a(t-t_3) - V_c(t-t_3) = L_x\left(\frac{d}{dt}I_{xa}(t-t_3) - \frac{d}{dt}I_{xc}(t-t_3)\right). \end{cases} \tag{3.10}$$

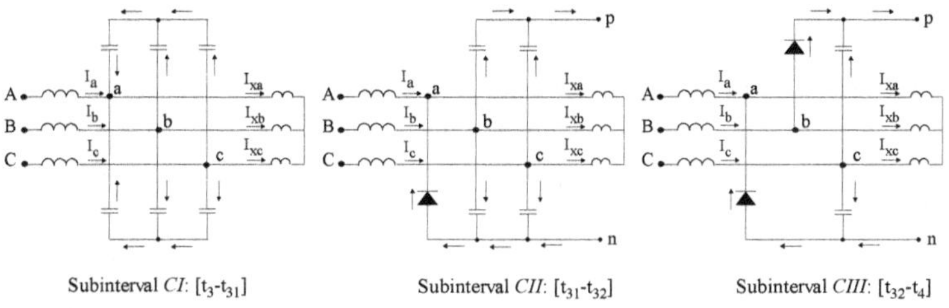

Fig. 3.3: Topological sequence of the resonant states (ACS-HW-ARC circuit).

By applying the Kirchhoff's current law to nodes a, b and c it results:

$$\begin{cases} I_a - 2C_S\left(\frac{d}{dt}V_a(t-t_3)\right) - I_{xa}(t-t_3) = 0, \\ I_b - 2C_S\left(\frac{d}{dt}V_c(t-t_3)\right) - I_{xb}(t-t_3) = 0, \\ I_c - 2C_S\left(\frac{d}{dt}V_c(t-t_3)\right) - I_{xb}(t-t_3) = 0. \end{cases} \tag{3.11}$$

Note, that $dV_a/dt < 0$. Introducing the auxiliary currents from (3.11) to (3.98):

$$\begin{cases} V_a(t-t_3) - V_b(t-t_3) = -\frac{1}{\omega_r^2}\left(\frac{d^2}{dt^2}V_a(t-t_3) - \frac{d^2}{dt^2}V_b(t-t_3)\right), \\ V_a(t-t_3) - V_c(t-t_3) = -\frac{1}{\omega_r^2}\left(\frac{d^2}{dt^2}V_a(t-t_3) - \frac{d^2}{dt^2}V_c(t-t_3)\right), \end{cases} \tag{3.12}$$

where ω_r is given by (3.1). The initial conditions at t_3 can be derived from (3.11) and (3.8) by substituting $2C_S = 1/(Z_r\omega_r)$ as follows:

$$\begin{cases} V_a(t_3) = \frac{V_{dc}}{2}, \quad V_b(t_3) = -\frac{V_{dc}}{2}, \quad V_c(t_3) = -\frac{V_{dc}}{2}, \\ \left.\frac{dV_b(t)}{dt}\right|_{t=t_3} = Z_r\,\omega_r\left(I_b - I_{xb}(t_3)\right) = Z_r\,\omega_r\left(I_b + \frac{1}{2}(I_a + I_{boost})\right), \\ \left.\frac{dV_c(t)}{dt}\right|_{t=t_3} = Z_r\,\omega_r\left(I_c - I_{xc}(t_3)\right) = Z_r\,\omega_r\left(I_c + \frac{1}{2}(I_a + I_{boost})\right). \end{cases} \tag{3.13}$$

The third differential equation required for deriving the expressions of the node voltages V_a, V_b and V_c can be obtained from equation, $I_{xa}(t)+I_{xb}(t)+I_{xc}(t)=0$, by replacing the auxiliary currents from (3.11) in this equation. It results to:

$$\frac{d}{dt}V_a(t-t_3)+\frac{d}{dt}V_b(t-t_3)+\frac{d}{dt}V_c(t-t_3)=0. \tag{3.14}$$

Solving the differential equation system formed by (3.12) and (3.14), the expressions describing the node voltages in the first subinterval of the resonant state, CI, are obtained as follows:

$$\begin{cases} V_a(t) &= -\frac{1}{6}V_{dc}-Z_r\,I_{boost}\,\sin\Big(\omega_r\,(t-t_3)\Big)+\frac{2}{3}V_{dc}\,\cos\Big(\omega_r\,(t-t_3)\Big), \\ V_b(t) &= -\frac{1}{6}V_{dc}+\frac{1}{2}Z_r\,(I_{boost}+I_b-I_c)\,\sin\Big(\omega_r\,(t-t_3)\Big)-\frac{1}{3}V_{dc}\,\cos\Big(\omega_r\,(t-t_3)\Big), \\ V_c(t) &= -\frac{1}{6}V_{dc}+\frac{1}{2}Z_r\,(I_{boost}+I_c-I_b)\,\sin\Big(\omega_r\,(t-t_3)\Big)-\frac{1}{3}V_{dc}\,\cos\Big(\omega_r\,(t-t_3)\Big). \end{cases} \tag{3.15}$$

Substituting $V_a(t)$, $V_b(t)$ and $V_c(t)$ from (3.15) in (3.11) yields:

$$\begin{cases} I_{xa}(t) &= I_a+\frac{2}{3}\frac{V_{dc}}{Z_r}\,\sin\Big(\omega_r\,(t-t_3)\Big)+I_{boost}\,\cos\Big(\omega_r\,(t-t_3)\Big), \\ I_{xb}(t) &= I_b-\frac{1}{3}\frac{V_{dc}}{Z_r}\,\sin\Big(\omega_r\,(t-t_3)\Big)-\frac{1}{2}\,(I_{boost}+I_b-Ic)\,\cos\Big(\omega_r\,(t-t_3)\Big), \\ I_{xb}(t) &= Ic-\frac{1}{3}\frac{V_{dc}}{Z_r}\,\sin\Big(\omega_r\,(t-t_3)\Big)-\frac{1}{2}\,(I_{boost}+I_c-Ib)\,\cos\Big(\omega_r\,(t-t_3)\Big). \end{cases} \tag{3.16}$$

Node voltage $V_a(t)$ has a larger $|dV/dt|$ at t_3 and therefore reaches the opposite DC rail voltage faster than the other node voltages because $I_{xa}(t_3)=-I_{xb}(t_3)-I_{xc}(t_3)$ resulting in more initial energy (boost energy) in the auxiliary inductor L_{xa}.

Assume at $t=t_{31}$, $V_a(t_{31})=-V_{dc}/2$ (see Fig. 2.9). Substituting $V_a(t_{31})$ from (3.15) and solving it for t_{31}:

$$t_{31} = \frac{2}{\omega_r} \arctan(\sqrt{3}\sqrt{3\,\rho^2+1} - 3\,\rho), \tag{3.17}$$

where

$$\rho = \frac{Z_r\, I_{boost}}{V_{dc}}. \tag{3.18}$$

Note, that ρ is the normalized boost current when V_{dc} and Z_r are assumed as the base values. At t_{31}, diode D_{an} turns on and the second subinterval of the resonant state, CII, begins.

During the second subinterval, $[t_{31} - t_{32}]$, the voltage at node a, $V_a(t)$, is clamped to the negative DC rail, while the other node voltages, $V_b(t)$ and $V_c(t)$, continue resonating to the positive DC rail. The initial conditions and the differential equations describing the time variation of $V_b(t)$ and $V_c(t)$ can be obtained in a similar way, as described for the first subinterval, as follows:

$$\begin{cases} V_b(t-t_{31}) &= -\dfrac{V_{dc}}{2} - \dfrac{1}{\omega_r^2}\left(2\,\dfrac{d^2}{dt^2}\,V_b(t-t_{31}) + \dfrac{d^2}{dt^2}V_c(t-t_{31})\right), \\ V_c(t-t_{31}) &= -\dfrac{V_{dc}}{2} - \dfrac{1}{\omega_r^2}\left(\dfrac{d^2}{dt^2}\,V_b(t-t_{31}) + 2\,\dfrac{d^2}{dt^2}V_c(t-t_{31})\right), \end{cases} \tag{3.19}$$

$$\begin{cases} V_b(t_{31}), \quad V_c(t_{31}) \qquad \text{from equation 3.15,} \\ \left.\dfrac{d\,V_b(t)}{dt}\right|_{t=t_{31}} = Z_r\,\omega_r\left[I_b - I_{xb}(t_{31})\right], \\ \left.\dfrac{d\,V_c(t)}{dt}\right|_{t=t_{31}} = Z_r\,\omega_r\left[I_c - I_{xc}(t_{31})\right]. \end{cases} \tag{3.20}$$

Despite the complexity of foregoing differential equations, it is still possible to solve these differential equations using computer algebra programs such as "Maple V". The solution will contain *sin*- and *cos*-terms of the both $\omega_{r1} = \omega_r$ and $\omega_{r2} = \dfrac{\omega_r}{\sqrt{3}}$ frequencies as follows:

$$\begin{cases} V_b(t) &= -\frac{V_{dc}}{2} + \Big\{K_{\omega_{r1}} \sin\Big(\omega_{r1}(t-t_{31})\Big) + K_b \cos\Big(\omega_{r1}(t-t_{31})\Big)\Big\}(I_b - I_c) \\ & \qquad + \frac{V_{dc}}{2}\Big\{K_{\omega_{r2}} \sin\Big(\omega_{r2}(t-t_{31}) + \cos\Big(\omega_{r2}(t-t_{31})\Big)\Big\}, \\ V_c(t) &= -\frac{V_{dc}}{2} + \Big\{K_{\omega_{r1}} \sin\Big(\omega_{r1}(t-t_{31})\Big) + K_c \cos\Big(\omega_{r1}(t-t_{31})\Big)\Big\}(I_c - I_b) \\ & \qquad + \frac{V_{dc}}{2}\Big\{K_{\omega_{r2}} \sin\Big(\omega_{r2}(t-t_{31}) + \cos\Big(\omega_{r2}(t-t_{31})\Big)\Big\}, \end{cases} \tag{3.21}$$

with

$$\begin{aligned} K_{\omega_{r1}} &= \frac{Z_r(2 - 3\sqrt{3}\sqrt{3\rho^2+1})}{2(9\rho^2+4)}, \\ K_{\omega_{r2}} &= \sqrt{3\rho^2+1}, \\ K_b &= \frac{Z_r(2\sqrt{3}\sqrt{3\rho^2+1} - 3\rho)}{2(9\rho^2+4)}, \\ K_c &= -\frac{Z_r(2\sqrt{3}\sqrt{3\rho^2+1} + 3\rho)}{2(9\rho^2+4)}, \end{aligned}$$

where ρ is defined by 3.18.

From (3.11) the expressions of the auxiliary currents for this part are calculated and given by:

$$\begin{cases} I_{xb}(t) &= I_b + \frac{1}{Z_r}\Big\{\Big[K_b \sin\Big(\omega_{r1}(t-t_{31})\Big) - K_{\omega_{r1}} \cos\Big(\omega_{r1}(t-t_{31})\Big)\Big](I_b - I_c) \\ & \qquad + \frac{\sqrt{3}V_{dc}}{6}\Big[\sin\Big(\omega_{r2}(t-t_{31})\Big) - K_{\omega_{r2}} \cos\Big(\omega_{r2}(t-t_{31})\Big)\Big]\Big\}, \\ I_{xc}(t) &= I_c + \frac{1}{Z_r}\Big\{\Big[-K_c \sin\Big(\omega_{r1}(t-t_{31})\Big) + K_{\omega_{r1}} \cos\Big(\omega_{r1}(t-t_{31})\Big)\Big](I_c - I_b) \\ & \qquad + \frac{\sqrt{3}V_{dc}}{6}\Big[\sin\Big(\omega_{r2}(t-t_{31})\Big) - K_{\omega_{r2}} \cos\Big(\omega_{r2}(t-t_{31})\Big)\Big]\Big\}, \\ I_{xa}(t) &= I_a + \frac{1}{Z_r}\Big\{\Big[-(K_b + K_c)\sin\Big(\omega_{r1}(t-t_{31})\Big) - 2K_{\omega_{r1}} \cos\Big(\omega_{r1}(t-t_{31})\Big)\Big] \\ & \qquad (I_b - I_c) + \frac{\sqrt{3}V_{dc}}{3}\Big[-\sin\Big(\omega_{r2}(t-t_{31})\Big) + K_{\omega_{r2}} \cos\Big(\omega_{r2}(t-t_{31})\Big)\Big]\Big\}. \end{cases} \tag{3.22}$$

Since $|I_b| < |I_c|$, the initial resonant energy of L_{xb} is higher than that of L_{xc}. This causes $V_b(t)$ to reach the positive DC rail voltage faster than $V_c(t)$.

The third subinterval of the resonant state begins when $V_b(t)$ equals $V_{dc}/2$ at t_{32} and diode D_{bp} begins conducting (see Fig. 3.5). This time can be derived from:

$$V_b(t_{32}) - \frac{V_{dc}}{2} = 0. \tag{3.23}$$

Substitution of $V_b(t_{32})$ from (3.21) in (3.23) results in a complex equation which is very difficult to solve parametrically.

Once the time t_{32} is calculated from (3.23) for the certain circuit characteristics (Z_r and ω_r) and load conditions, the initial conditions for the third subinterval can be also obtained as follows:

$$\begin{cases} V_c(t_{32}) & \text{from equation 3.21,} \\ \left.\dfrac{d\,V_c(t)}{dt}\right|_{t=t_{32}} & = Z_r\,\omega_r \Big[I_c - I_{xc}(t_{32})\Big]. \end{cases} \tag{3.24}$$

The differential equation describing the variation of $V_c(t)$ during the third subinterval is given by:

$$V_c(t - t_{32}) = -\frac{3}{2}\frac{1}{\omega_r^2}\frac{d^2}{dt^2}V_c(t - t_{32}). \tag{3.25}$$

After solving this equation, an expression containing *sin*- and *cos*-terms of the frequency $\omega_{r3} = \sqrt{\frac{2}{3}}\,\omega_r$ is obtained as follows:

$$V_c(t) = K_{1\omega_{r3}}\, sin\Big(\omega_{r3}\,(t - t_{32})\Big) + K_{2\omega_{r3}}\, cos\Big(\omega_{r3}\,(t - t_{32})\Big), \tag{3.26}$$

where $K_{1\omega_{r3}}$, $K_{2\omega_{r3}}$ are functions of I_a, I_b, I_c, I_{boost}, V_{dc}, L_x and C_s.

Once the expression of $V_c(t)$ is also known, the auxiliary current equations can be derived by the sequentially solving of the following set of equations.

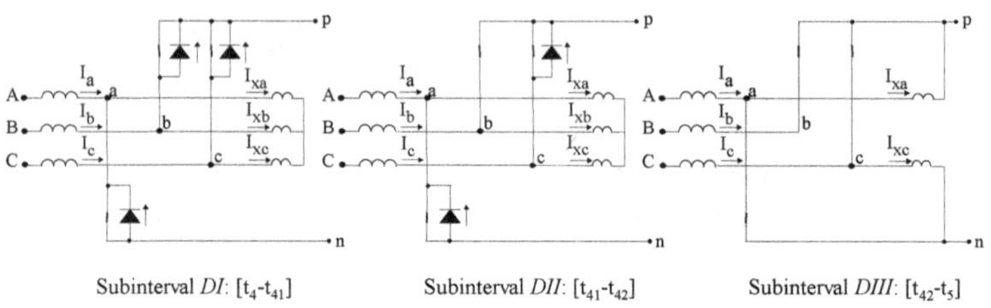

Fig. 3.4: Topological sequence of the discharging state (ACS-HW-ARC circuit)

$$\begin{cases} I_{xc}(t-t_{32}) &= I_c - Z_r\,\omega_r\,\dfrac{d\,V_c(t-t_{32})}{dt}, \\ \dfrac{d\,I_{xb}(t-t_{32})}{dt} &= \dfrac{V_{dc}\,Z_r}{2\,\omega_r} - \dfrac{1}{2}\,\dfrac{d\,I_{xc}(t-t_{32})}{dt} \quad \{I_{xb}(t_{32}) \text{ from Eq. 3.22}\}, \\ I_{xa}(t-t_{32}) &= -I_{xb}(t-t_{32}) - I_{xc}(t-t_{32}). \end{cases} \tag{3.27}$$

Finally at $t = t_4$ the node voltage $V_c(t)$ also reaches the positive DC rail voltage and the resonant state ends. This time can be derived from:

$$V_c(t_4) - \frac{V_{dc}}{2} = 0. \tag{3.28}$$

Total duration of the resonant state is defined by $\Delta t_{res} = t_4 - t_3$.

c) Discharging State:

At the end of the resonant state, the diodes D_{an}, D_{bp} and D_{cp} conduct and provide the ZVS conditions for their antiparallel switches S_{an}, S_{bp} and S_{cp}. The switches are turned on simultaneously at t_4 and the discharging state starts. The discharging process is also divided into the three subintervals as shown in Fig. 3.4.

The antiparallel diodes conduct as long as the respective auxiliary currents are larger than the phase currents (subinterval DI). When $|I_{xb}|$ becomes smaller than $|I_b|$ at t_{41}, diode D_{bp} turns off and the switch S_{bp} conducts the current difference $|I_b - I_{xb}|$ (subinterval DII). At t_{42}, the end of subinterval DII, the auxiliary current I_{xb} becomes zero and the auxiliary switch S_x is turned off. At t_5, the end of subinterval $DIII$, I_{xa} and I_{xc} also become zero and the auxiliary circuit action ends. The times t_{41}, t_{42} and t_5 are given by:

$$t_{41} = 3\,\frac{L_x}{V_{dc}}\,\Big[|I_{xb}(t_4) - I_b|\Big] + t_4, \tag{3.29}$$

$$t_{42} = 3\,\frac{L_x}{V_{dc}}\,|I_{xb}(t_4)| + t_4, \tag{3.30}$$

$$t_5 = t_{42} + \frac{L_x}{V_{dc}}\,\Big[I_{xa}(t_4) - 2I_{xb}(t_4)\Big] = \frac{L_x}{V_{dc}}\,\Big[I_{xa}(t_4) + I_{xb}(t_4)\Big] + t_4, \tag{3.31}$$

where $I_{xb}(t_4)$ and $I_{xa}(t_4)$ are the auxiliary currents at the end of the resonant state obtained from (3.27). Total duration of the discharging state is defined by $\Delta t_{dis} = t_5 - t_4$.

For verification of derived expressions simulations are performed with ideal MOSFET model of SABER simulator [38]. Both the calculative and simulative results are illustrated in Fig. 3.5. The agreement of simulated waveforms with the calculated ones verifies the expressions obtained in this section.

3.1.1.2 Determination of the Boost Energy

Because of the topological change of the auxiliary network during the resonant state as shown in Fig. 3.3, the achievable maximum value of V_c in (3.26) falls to a value smaller than $V_{dc}/2$, if I_{boost} is not large enough. This maximum value can be derived from (3.26) as follows: the time $t_{V_{c,max}}$, at which the node voltage V_c reaches its maximum value $V_{c,max}$ in case of absence of diode D_{cp}, can be obtained by differentiating $V_c(t)$ in (3.26) as follows:

$$\frac{d\,V_c(t)}{dt} = K_{1\omega_{r3}}\,\omega_{r3}\,\cos\Big(\omega_{r3}\,(t - t_{32})\Big) - K_{2\omega_{r3}}\,\omega_{r3}\,\sin\Big(\omega_{r3}\,(t - t_{32})\Big) = 0. \tag{3.32}$$

Solving this equation for t gives:

$$t_{V_{c,max}} = t_{32} + \frac{1}{\omega_{r3}}\,\arctan\left(\frac{K_{1\omega_{r3}}}{K_{2\omega_{r3}}}\right). \tag{3.33}$$

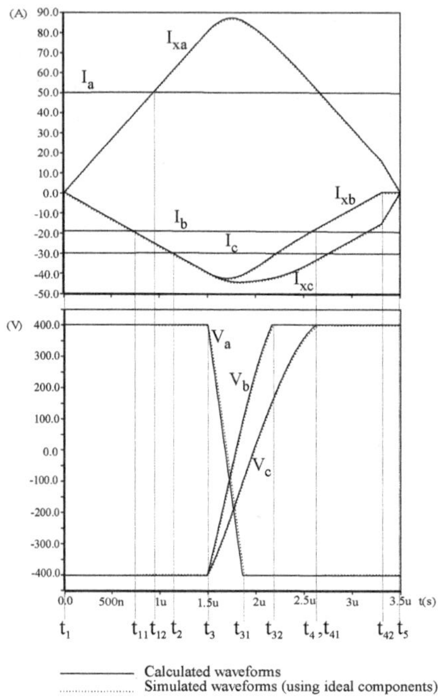

Fig. 3.5: Calculative and simulative operation waveforms for the ACS-HW-ARC circuit ($Z_r = 24.7\Omega$ and $\omega_r = 2469324\,rad/s$).

Substitution of it into (3.26) results in $V_{c,max}$:

$$V_{c,max} = \sqrt{K_{1\omega_{r3}}^2 + K_{2\omega_{r3}}^2}. \tag{3.34}$$

This maximum value is smaller than $V_{dc}/2$, if I_{boost} is not large enough. In such case the ZVS conditions cannot be achieved completely. Fig. 3.6 shows the results of the simulation of two situations: on the left side the completely achieved ZVS conditions for all phases and on the right side the uncompleted ZVS conditions for phases b and c due to a lack of boost energy. In this Figure, the continuation of V_c is also drawn by a dotted line in case of free resonating, when the diode

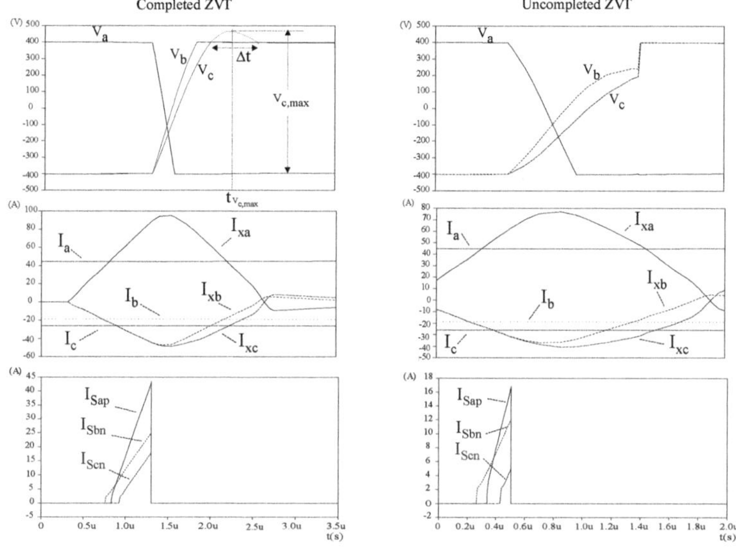

Fig. 3.6: Completed and uncompleted ZVS conditions

D_{cp} does not exist.

Boost energy should be high enough in order to increase $V_{c,max}$ to a value beyond $V_{dc}/2$ to ensure the conduction of diode D_{cp} for a hold-on duration Δt as shown in Fig. 3.6(left). The hold-on duration determined by the switch turn-on delay time guarantees the ZVS turn-on for S_{cp}. The necessary boost energy (I_{boost}) can be determined for a certain load condition by numerically solving the following equation:

$$\Delta t = |t_{V_{c1}} - t_{V_{c2}}|, \tag{3.35}$$

where $t_{V_{c1}}$ and $t_{V_{c2}}$ are the crossing points of waveform V_c with the positive DC link voltage, $V_{dc}/2$, and can be obtained by solving the equation $V_c(t) - V_{dc}/2 = 0$. It results in:

$$t_{v_{c1}}, t_{v_{c2}} = t_{32} + \frac{2}{\omega_{r3}} \arctan\left(\frac{2\,K_{1\omega_{r3}} \pm \sqrt{4\,(K^2_{1\omega_{r3}} + K^2_{2\omega_{r3}}) - V^2_{dc}}}{V_{dc} + 2\,K_{2\omega_{r3}}}\right). \quad (3.36)$$

Note, that parameters $K_{1\omega_{r3}}$ and $K_{2\omega_{r3}}$ depend on I_{boost}.

The circuit is assumed to be ideal in the above calculation, so the resulting I_{boost} has to be increased further by some percent to cover the conduction loss of the auxiliary devices.

For simplicity of the control circuit it is mostly desirable to have a constant charging duration for convenience. Under this condition, if the charging duration including the duration of boost charging is set, so that the auxiliary current I_{xa} reaches a value at least $\sqrt{3}$ times the amplitude of the phase current at full load, a complete ZVS can be always achieved [37]. This value is determined considering this fact that all auxiliary currents have to exceed their respective phase currents at the end of the charging state. At worst case, when $\phi = 30^o$, $I_b = 0$ and $I_a = -I_c = \sqrt{3}\,\hat{I}_{Ph}/2$, the auxiliary current I_{xa} has to increase by $\sqrt{3}$ times the amplitude of phase current $\hat{I}_{Ph}$, thereby I_{xc} can exceed I_c. (Note, that during the charging state the relation $I_{xa} = 2\,|I_{xb}| = 2|\,I_{xc}|$ holds.)

If a lower value for the auxiliary current is chosen, ZVS is partially lost during parts of the AC line cycle, resulting in increase of switching losses and EMI. Taking into account the increase of the auxiliary inductor currents during the resonant state, and adding some margin, it results that the auxiliary switch S_x should be selected to withstand a peak current two times larger than the maximum amplitude of the phase currents.

3.1.1.3 Device Stresses

During the ZVS process, because of the larger boost energy in the auxiliary inductor of phase a, the maximum dV/dt appears at this phase. $(dV/dt)_{max}$ can be obtained from (3.15):

$$\left(\frac{dV}{dt}\right)_{max} = \frac{1}{6}\,\omega_r\,V_{dc}\sqrt{\left(\frac{3\,\rho}{2}\right)^2 + 1}, \quad (3.37)$$

which occurs at:

$$t_{max} = \frac{1}{\omega_r} \arctan\left(\frac{2}{3\,\rho}\right), \tag{3.38}$$

where ρ given by (3.18). $(dI/dt)_{max}$ is determined by the auxiliary inductors and given by:

$$\left(\frac{dI}{dt}\right)_{max} = \frac{2}{3}\frac{V_{dc}}{L_x}. \tag{3.39}$$

At t_{max}, the auxiliary switch current also reaches its maximum value given by:

$$I_{S_x,max} = I_a + \frac{2\,V_{dc}}{3\,Z_r}\sqrt{\left(\frac{3\,\rho}{2}\right)^2 + 1}. \tag{3.40}$$

3.1.2 The ACS-HW-ARCP Circuit

The ACS-HW-ARCP circuit is illustrated along with the switching state sequence in Fig. 3.7. The analysis is performed for the transition between switching states $S_{1(pnn)}$ and $S_{7(ppp)}$, the same as used for describing the operation of circuit in section 2.3.2.1 of chapter 2. For the operation waveforms of circuit see Fig. 2.13 and Fig. 3.11.

3.1.2.1 Analysis

a) Charging and Boost Charging State:

Topological sequence of charging (interval A) and boost charging (interval B) states of this circuit are shown in Fig. 3.8. The charging state starts with the turn-on of switches $S_{xb(n)}$ and $S_{xc(n)}$ at $t_1 = 0$. Thereby half of V_{dc} is applied across the auxiliary inductor L_x.

The increase of the auxiliary inductor current is described by:

$$I_{L_x}(t) = -\frac{1}{2}\frac{V_{dc}}{L_x}\,t. \tag{3.41}$$

In an ideal circuit, the auxiliary inductor current I_{L_x} is equally divided between auxiliary currents I_{xb} and I_{xc} but this is far from being practical. Because

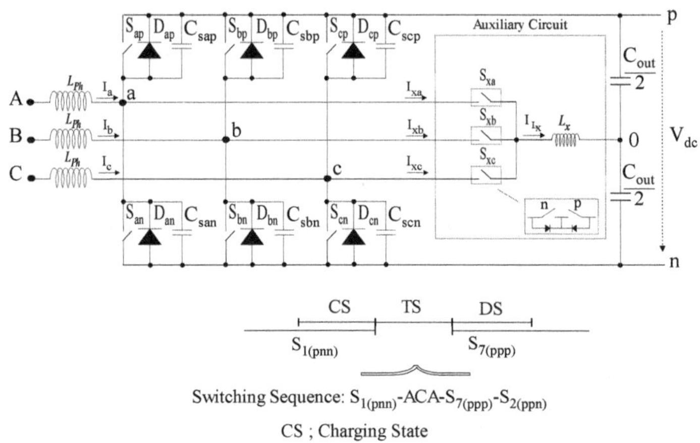

Fig. 3.7: ASC-HW-ARCP PWM converter and the switching state sequence in the first half of sector I

the division of I_{L_x} into I_{xb} and I_{xc} is determined by the voltage difference between the nodes b and c caused by differing the current levels of I_b and I_c, the voltage drop across the main switches and diodes should be taken into consideration. Considering only the resistance of the auxiliary branch r_x and of the main branch r_m at phases b and c, the charging stage can be described as follows: since $|I_b| < |I_c|$, the absolute value of V_b is smaller than that of V_c. Due to this voltage difference, after turning on $S_{xb(n)}$ and $S_{xc(n)}$ the auxiliary current I_{xc} increases very faster than I_{xb} till the equilibrium point of the current division described by the following equation is achieved:

$$I_{xb}(t) - I_{xc}(t) = \frac{r_m}{r_m + r_x}(I_b - I_c). \tag{3.42}$$

Thereafter both I_{xb} and I_{xc} increase nearly with the same dI/dt equal to half of dI/dt of I_{L_x}. When I_{xb} reaches I_b and D_{bn} turns off, I_{L_x} is divided no longer

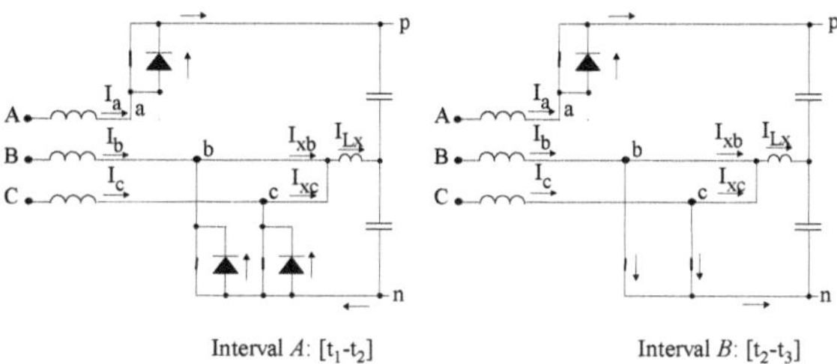

Fig. 3.8: Topological sequence of charging and boost charging states for the ACS-HW-ARCP circuit

according to (3.42). I_{xb} nearly remains constant while I_{xc} increases with a dI/dt almost equal to that of I_{L_x} (see Fig. 2.13(a)). At t_2, the end of charging state, when I_{L_x} exceeds the value $(I_b + I_c)$, the diode D_{cn} also turns off. The auxiliary inductor L_x continues charging for the required boost energy. The surplus current flows through the auxiliary branch of phase b[1] because of the different voltage drop across the auxiliary branches. At t_3, the end of boost charging state, when the magnitude of I_{L_x} reaches the value $(|I_b| + |I_c| + I_{boost})$, S_{bn} and S_{cn} are simultaneously turned off, the circuit is transferred to the resonant state and the capacitances paralleled to the switches take effect. The times t_2 and t_3 are give by:

$$t_2 = \frac{2\,L_x\,(|I_b| + |I_c|)}{V_{dc}}, \tag{3.43}$$

$$t_3 = \frac{2\,L_x\,(|I_b| + |I_c| + I_{boost})}{V_{dc}}. \tag{3.44}$$

The total duration of charging state yields as:

$$\varDelta t_{ch} = t_3 - t_1 = \frac{2\,L_x\,(|I_b| + |I_c| + I_{boost})}{V_{dc}}, \tag{3.45}$$

which is given for the simultaneous commutation of two phase currents. For commutation of only one phase current it is given by:

[1] Assuming $(|I_b| + I_{boost}) < |I_c|$

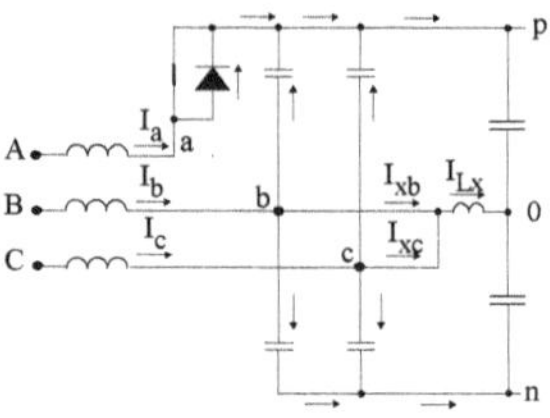

Fig. 3.9: ACS-HW-ARCP circuit during the resonant state.

$$\Delta t_{ch} = t_3 - t_1 = \frac{2\,L_x\,(|I_{ph}| + I_{boost})}{V_{dc}}. \tag{3.46}$$

B) Resonant state:

The turn-off of the switches S_{bn} and S_{cn}, divides the circuit into two similar resonant circuits sharing a common resonant inductor L_x (see Fig. 3.9). This leads to an equal division of the boost current, I_{boost}, between the two equal resonant circuits (see Fig. 2.13(a)). The auxiliary currents I_{xb} and I_{xc} at the beginning of the resonant state can be then given by:

$$\begin{cases} I_{L_x}(t_3) &=& -(|I_b| + |I_c| + I_{boost}), \\ I_{bx}(t_3) &=& -\left(|I_b| + \dfrac{I_{boost}}{2}\right), \\ I_{cx}(t_3) &=& -\left(|I_c| + \dfrac{I_{boost}}{2}\right). \end{cases} \tag{3.47}$$

During the resonant state, the variations of V_b and V_c are the same and can be described by the following differential equation:

$$V_{b,c}(t - t_3) = -\frac{K}{\omega_r^2}\,\frac{d^2}{dt^2}\,V_{b,c}(t - t_3), \tag{3.48}$$

where K is the number of phase currents being commutated ($K = 2$ is selected in this example).

Referring to (3.11) and (3.47), the initial conditions can be given as follows:

$$\begin{cases} V_{b,c}(t_3) = -\dfrac{V_{dc}}{2}, \\ \dfrac{d\,V_{b,c}(t)}{dt}\bigg|_{t=t_3} = \dfrac{Z_r\,\omega_r}{2}\,I_{boost}. \end{cases} \tag{3.49}$$

After solving this differential equation, the following expression describing the variation of node voltages $V_b(t)$ and $V_c(t)$ is obtained:

$$V_{b,c}(t) = \frac{V_{dc}}{2}\left\{\sqrt{2}\,\rho\,\sin\left(\frac{\omega_r}{\sqrt{2}}\,(t-t_3)\right) - \cos\left(\frac{\omega_r}{\sqrt{2}}\,(t-t_3)\right)\right\}, \tag{3.50}$$

where ρ is given by (3.18). Substituting (3.50) in (3.11) results in:

$$\begin{cases} I_{xb,xc}(t) = I_{b,c} - \dfrac{V_{dc}}{4\,Z_r}\left\{2\,\rho\,\cos\left(\dfrac{\omega_r}{\sqrt{2}}\,(t-t_3)\right) + \sqrt{2}\,\sin\left(\dfrac{\omega_r}{\sqrt{2}}\,(t-t_3)\right)\right\}, \\ I_{L_x}(t) = I_{xb}(t) + I_{xc}(t). \end{cases} \tag{3.51}$$

Replacing the left side of equation (3.50) by $V_{dc}/2$ and solving it for t, results in time t_4, at which the node voltages reach the positive DC rail voltage and the resonant state ends:

$$t_4 = t_3 + \frac{2\,\sqrt{2}}{\omega_r}\,\arctan\left(\frac{1}{\sqrt{2}\,\rho}\right). \tag{3.52}$$

Hence, the duration of resonant state Δt_{res} yields as:

$$\Delta t_{res} = t_4 - t_3 = \frac{2\,\sqrt{2}}{\omega_r}\,\arctan\left(\frac{1}{\sqrt{2}\,\rho}\right). \tag{3.53}$$

c) Discharging State:

The auxiliary circuit operates during the discharging state like the charging state, but in a reverse sequence. When $|I_{L_x}|$ becomes smaller than the value $|(I_b + I_c)|$ at t_{41}, the switches S_{bp} and S_{cp} start conducting. At this time the first subinterval of the discharging state is concluded as shown in Fig. 3.10. It is calculated by:

$$t_{41} = t_4 + \Delta t = t_4 + \frac{2\,L_x\,[|I_{L_x}(t_4) - (I_b + I_c)|]}{V_{dc}}, \tag{3.54}$$

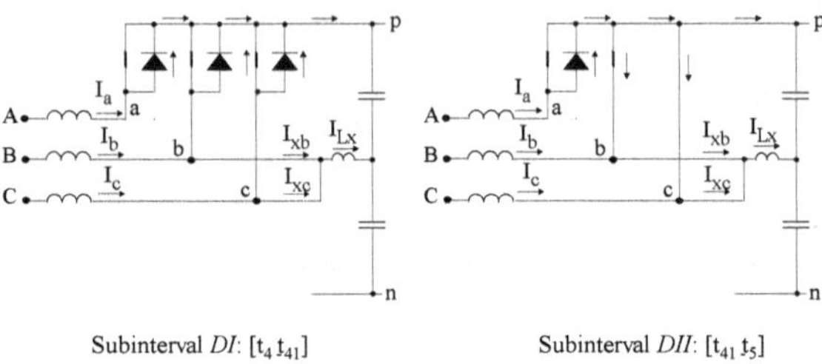

Fig. 3.10: ACS-HW-ARCP circuit during the discharging state

where Δt represents the necessary conduction duration of diodes D_{bp} and D_{cp}, ensuring the ZVS conditions.

At t_5, the end of the discharging state, the auxiliary inductor current I_{L_x} decays to zero and the auxiliary switches $S_{xb(n)}$ and $S_{xc(n)}$ can be turned off under zero current conditions. The time t_5 and the duration of the discharging state for the simultaneous commutation of two phase currents is obtained by:

$$t_5 = t_4 + \frac{2\,L_x\,|I_{L_x}(t_4)|}{V_{dc}} = t_3 + \Delta t + \frac{2\,L_x\,(|I_b + I_C|)}{V_{dc}}, \tag{3.55}$$

$$\Delta t_{dis} = t_5 - t_4 = \Delta t + \frac{2\,L_x\,(|I_b + I_c|)}{V_{dc}}. \tag{3.56}$$

For the commutation of only one phase current Δt_{dis} results:

$$\Delta t_{dis} = t_4 - t_3 = \Delta t + \frac{2\,L_x\,(|I_{ph}|)}{V_{dc}}. \tag{3.57}$$

Fig. 3.11 shows the operation waveforms of the converter resulting from the derived expressions along with the simulated ones during one ACA. In the simulations saturation voltage and on-state resistance of the switches, $V_{sat(S)}$ and r_S, as well as junction voltage and dynamic resistance of the diodes, $V_{j(D)}$ and r_D, are considered. Both the calculated and simulated waveforms are obtained with $I_{boost} = 0$. As it can be clearly seen from the simulated waveforms, without boost energy, node voltage V_b and V_c cannot exactly reach the positive DC rail

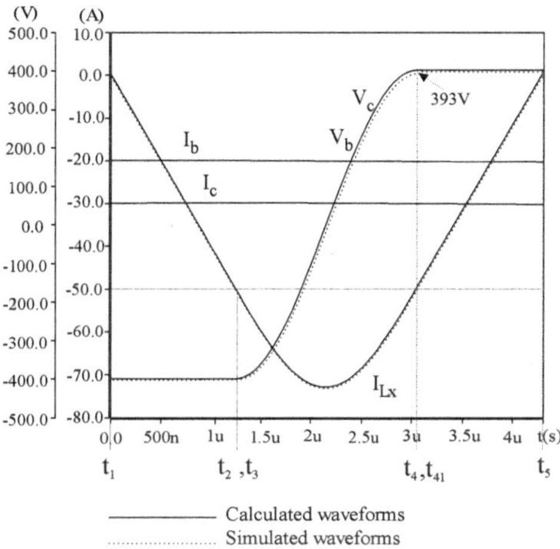

Fig. 3.11: Operation waveforms of the ACS-HW-ARCP circuit during one ACA with $I_{boost} = 0$. ($Z_r = 24.7\Omega$, $\omega_r = 2469324\,rad/s$ and in the simulation models $V_{sat(S)} = 2V$, $r_S = 40m\Omega$ $V_{j(D)} = 0.7V$ and $r_D = 40m\Omega$ are assumed.)

voltage because of the voltage drop and conduction loss of the auxiliary devices. Comparison of the calculated waveforms with simulated ones shown in this figure and also with those shown in Fig. 2.13(a) verifies the derived expressions.

3.1.2.2 Determination of the Boost Energy

In a real circuit, the node voltages have to resonate to a value higher than the DC rail voltage, for the antiparallel diodes D_{bp} and D_{cp} start conducting and remain conducting for the duration Δt till the turn-on transition of the switches S_{bp} and S_{cp} ends. Considering this condition, the necessary boost energy (I_{boost}) can be calculated from:

$$\Delta t = \frac{T'}{2} - \Delta t_{res} = \frac{\sqrt{2}}{\omega_r}\left\{\pi - 2\ \arctan\left(\frac{1}{\sqrt{2}\,\rho}\right)\right\}, \tag{3.58}$$

with $T' = \dfrac{2\,\pi}{\omega_r'}$ and $\omega_r' = \dfrac{\omega_r}{\sqrt{2}}$.

After simplifying the above given equation, it results in:

$$I_{boost(\Delta t)} = \frac{V_{dc}}{\sqrt{2}\, Z_r} \tan\left(\frac{\omega_r\, \Delta t}{2\sqrt{2}}\right). \tag{3.59}$$

This boost energy would ensure the duration Δt for ideal auxiliary devices. Considering conduction loss of the auxiliary devices, additional boost energy is necessary for compensating this loss. So the total necessary boost current can be good approximated by:

$$I_{boost} = I_{boost(\Delta t)} + \sqrt{\frac{\Delta t_{res}}{L_x}\, 2\, I_{S_x,max}\, (V_{CE(on)} + V_{FM})}, \tag{3.60}$$

where $V_{CE(on)}$ and V_{FM} are collector-to-emitter saturation voltage of the auxiliary switches and forward voltage drop of their antiparallel diode respectively and $I_{S_x,max}$ is the maximum current of the auxiliary switch.

3.1.2.3 Device Stresses

The maximum dV/dt on the main switches can be calculated by differentiating (3.50), leading to:

$$\left(\frac{dV}{dt}\right)_{max} = \frac{\sqrt{2}}{4}\, \omega_r\, V_{dc}\, \sqrt{2\,\rho^2 + 1}. \tag{3.61}$$

This occurs at:

$$t_{max} = \frac{\sqrt{2}}{\omega_r} \arctan\left(\frac{1}{\sqrt{2}\,\rho}\right). \tag{3.62}$$

At this time the currents of the auxiliary switches also attain their maximum values given by:

$$I_{S_{xb,\,xc},max} = |I_{b,c}| + \frac{1}{2\sqrt{2}} \frac{V_{dc}}{Z_r} \sqrt{2\,\rho^2 + 1}. \tag{3.63}$$

$(dI/dt)_{max}$ of the main switches in simultaneous commutation of two phase currents occurs during the boost charging and is equal to $(dI_{L_x}/dt)_{max}$ given by:

$$\left(\frac{dI_{L_x}}{dt}\right)_{max} = \frac{V_{dc}}{2\, L_x}. \tag{3.64}$$

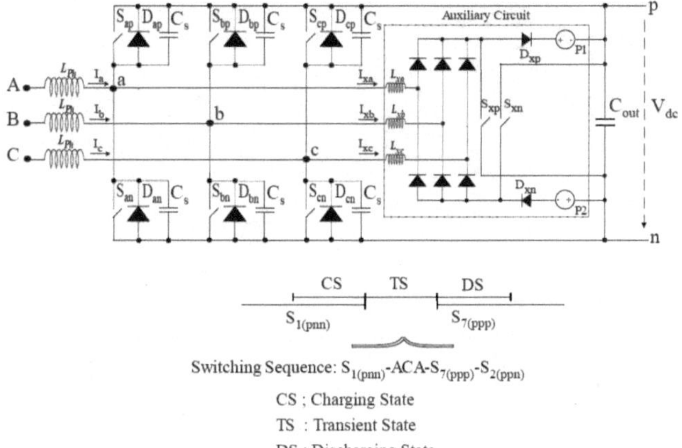

Fig. 3.12: Simplified ASC-QW-ARC PWM converter and the switching state sequence in the first half of sector I.

3.1.3 The ACS-QW-ARC Circuit

The ACS-QW-ARC circuit and the switching state sequence are shown in Fig. 3.12. The circuit will be analysed for the same transition $S_{1(pnn)} \rightarrow S_{7(ppp)}$ as used for the description of the circuit operation in chapter 2. See Fig. 3.15(a) for the simulative waveforms of converter operation (See also Fig. 2.16(a)).

3.1.3.1 Analysis

a) Charging and Resonant States:

The charging state begins when the auxiliary switch S_{xn} is turned on at $t_1 = 0$. The DC link voltage V_{dc} is applied on the auxiliary inductors L_{xb} and L_{xc} causing the auxiliary currents I_{xb} and I_{xc} to increase linearly. See Fig. 3.13 for the topological sequence of the circuit.

At t_{2b} and t_{2c} respectively, the auxiliary currents I_{xb} and I_{xc} reach the phase currents I_b and I_c, whereby diodes D_{bn} and D_{cn} are turned off and the resonant state in phases b and c begins. The times t_{2b} and t_{2c} are obtained from:

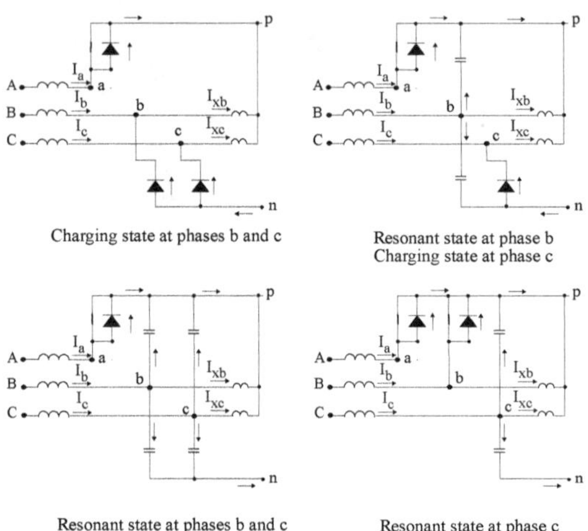

Fig. 3.13: Topological sequence of the simplified ASC-QW-ARC PWM converter during the charging and resonant states.

$$\begin{cases} t_{2b} &= \dfrac{L_x\,|I_b|}{V_{dc}}, \\ t_{2c} &= \dfrac{L_x\,|I_c|}{V_{dc}}. \end{cases} \tag{3.65}$$

Generally the duration of the charging phase is defined as $\Delta t_{ch} = \dfrac{L_x\,|I_{ph}|}{V_{dc}}$, where I_{ph} is the phase current.

Note, that the resonant state in each phase starts immediately after turn-off of the respective diode as described in section 2.3.3.1 of the last chapter. The circuit equations during the resonant state are given by:

$$V_{b,c}(t - t_{2b,2c}) + \frac{1}{\omega_r^2}\frac{d^2}{dt^2}V_{b,c}(t - t_{2b,2c}) = \frac{V_{dc}}{2}, \tag{3.66}$$

with the following initial conditions (refer to equation (3.11)):

$$\begin{cases} V_{b0} = V_b(t_{2b}) = -\dfrac{V_{dc}}{2}, \\ V_{c0} = V_c(t_{2c}) = -\dfrac{V_{dc}}{2}, \\ \left.\dfrac{d\,V_b(t)}{dt}\right|_{t=t_{2b}} = 0, \\ \left.\dfrac{d\,V_c(t)}{dt}\right|_{t=t_{2c}} = 0. \end{cases} \tag{3.67}$$

After solving the foregoing equations, the expressions describing the variation of the node voltages are obtained as follows:

$$\begin{cases} V_b(t) = V_{dc}\left[\dfrac{1}{2} - \cos\Big(\omega_r\,(t - t_{2b})\Big)\right], \\ V_c(t) = V_{dc}\left[\dfrac{1}{2} - \cos\Big(\omega_r\,(t - t_{2c})\Big)\right]. \end{cases} \tag{3.68}$$

The equations describing the auxiliary currents can also be derived by substituting (3.68) in (3.11). This results in:

$$\begin{cases} I_{xb}(t) = I_b - \dfrac{V_{dc}}{Z_r}\sin\Big(\omega_r\,(t - t_{2b})\Big), \\ I_{xc}(t) = I_c - \dfrac{V_{dc}}{Z_r}\sin\Big(\omega_r\,(t - t_{2c})\Big). \end{cases} \tag{3.69}$$

The resonant state ends for phases b and c at t_{3b} and t_{3c} respectively. Replacing the left side of the expressions given in (3.68) with $V_{dc}/2$ and solving them for t results in:

$$t_{3b,3c} = t_{2b,2c} + \frac{1}{2}\,\frac{\pi}{\omega_r}. \tag{3.70}$$

The duration of resonant state is given by $\Delta t_{res} = \dfrac{1}{2}\,\dfrac{\pi}{\omega_r}$ which is a quarter of the resonance period $T_r = \dfrac{2\,\pi}{\omega_r}$.

Since $t_{3b} < t_{3c}$, D_{bp} starts conducting sooner than D_{cp}. The auxiliary current I_{xb} remains constant until the auxiliary switch S_{xn} is turned off at t_{3c}. The switches S_{bp} and S_{cp} are also turned on at this time.

The auxiliary currents at the end of the resonant state t_{3c} are given by:

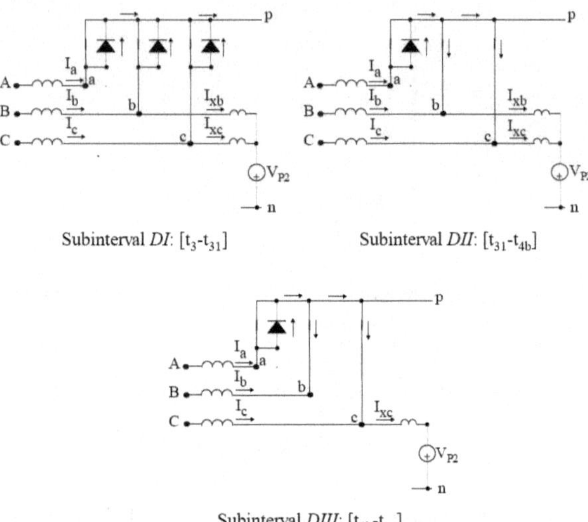

Fig. 3.14: Topological sequence of the simplified ASC-QW-ARC PWM converter during the discharging states

$$\begin{cases} I_{xb}(t_{3c}) & = & I_b - \dfrac{V_{dc}}{Z_r}, \\ I_{xc}(t_{3c}) & = & I_c - \dfrac{V_{dc}}{Z_r}. \end{cases} \tag{3.71}$$

b) Discharging State:

Discharging state starts after turning off the auxiliary switch S_{xn}. The auxiliary diode D_{xn} becomes conducting and the sum of V_{dc} and V_{P2} is applied across the auxiliary inductors causing the auxiliary current to decrease. This state is divided into three subintervals as shown in Fig. 3.14.

As the auxiliary currents decrease, at t_{31}, the end of the first subinterval DI, D_{bp} and D_{cp} turn off, and their antiparallel switch starts conducting. At t_{4b} and t_{4c}, the end of the discharging state at phases b and c respectively, the auxiliary currents I_{xb} and I_{xc} decay to zero. The times t_{31}, t_{4b} and t_{4c} are given by:

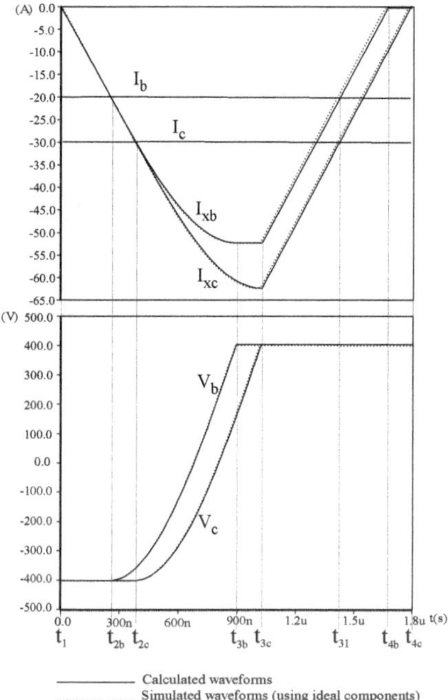

Fig. 3.15: Calculated and simulated operation waveforms of the ASC-QW-ARC PWM converter during one ACA ($Z_r = 24.7\Omega$, $\omega_r = 2469324\, rad/s$ and $V_P = 15V$).

$$\begin{cases} t_{31} &= t_{3c} + \dfrac{L_x V_{dc}}{Z_r (V_{dc} + V_{P2})} = t_{3c} + \dfrac{V_{dc}}{\omega_r (V_{dc} + V_{P2})}, \\ t_{4b} &= t_{3c} + \dfrac{Z_r |I_b| + V_{dc}}{\omega_r (V_{dc} + V_{P2})}, \\ t_{4c} &= t_{3c} + \dfrac{Z_r |I_c| + V_{dc}}{\omega_r (V_{dc} + V_{P2})}. \end{cases} \tag{3.72}$$

Consequently the duration of the discharging state is defined as $\Delta t_{dis} = \dfrac{Z_r |I_{ph}| + V_{dc}}{\omega_r (V_{dc} + V_P)}$.

The conformity of calculated waveforms with the simulated ones as shown in Fig. 3.15 confirms the analysis and derived expressions (See also Fig. 2.16(a)).

3.1.3.2 Device Stresses

The maximum dV/dt on the main switches is given by:

$$\left(\frac{dV}{dt}\right)_{max} = \omega_r V_{dc}. \tag{3.73}$$

This appears at $t_{max} = \frac{1}{2}\frac{\pi}{\omega_r}$.

The maximum dI/dt for the main switches is also given by:

$$\left(\frac{dI}{dt}\right)_{max} = \frac{V_{dc}}{L_x}. \tag{3.74}$$

The maximum value of the auxiliary switch current for simultaneous commutation of two phase currents is calculated from:

$$I_{S_{xn},max} = |I_b + I_c| + 2\frac{V_{dc}}{Z_r}, \tag{3.75}$$

and for commutation of only one phase current I_{ph}[2)] :

$$I_{S_{xn},max} = |I_{ph}| + \frac{V_{dc}}{Z_r}. \tag{3.76}$$

3.1.4 The ACS-QW-ARCP Circuit

The ACS-QW-ARCP circuit shown in Fig. 3.16 is analysed as before for the transition between the switching states $S_{1(pnn)}$ and $S_{7(ppp)}$.

Since the analysis of this circuit does not essentially differ from the analysis of ACS-QW-ARC circuit described in the previous section, only a short description containing the modified equations is given. The important point is that the commutation of the phase currents I_b and I_c are independent from each other, therefore only the commutation of I_b will be analysed, for the commutation processes in both phases are similar. For the simulative waveforms of converter operation see Fig. 2.21.

a) Commutation Processes:

---

2) The phase whose voltage polarity is opposite to polarity of others.

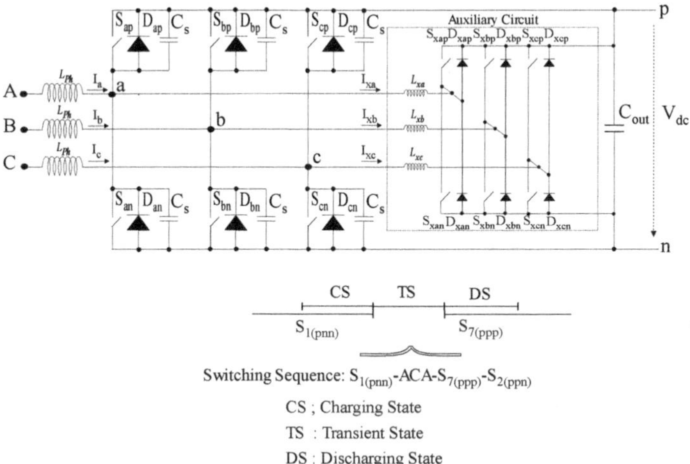

Fig. 3.16: The ACS-QW-ARCP PWM converter and the switching state sequence in the first half of sector I.

The commutation of I_b from D_{bn} to S_{bp} begins with turning on the auxiliary switch S_{xbp} at $t_1 = 0$, which brings the auxiliary circuit into the charging state. The three circuit states during the current commutation are shown in Fig. 3.17. The auxiliary current I_{xb} increases in the charging state as much as I_b. The end of the charging state, t_2, is given by:

$$t_2 = \frac{L_x \left| I_b \right|}{V_{dc}}. \tag{3.77}$$

The duration of the charging state is then defined as $\Delta t_{ch} = t_2 - t_1$.

At t_2, D_{bn} turns off and the resonant state begins. During the resonant state the equations given in (3.68) and (3.69) describe V_b and I_{xb} respectively. At the end of the resonant state t_3, given by (3.70), V_b reaches the positive DC rail. Switches S_{bp} and S_{xbp} are turned off, the diode D_{xbn} begins conducting and the circuit is brought into the discharging state. At t_4, given by the following equation, the auxiliary current I_{xb} reaches zero and the discharging state ends.

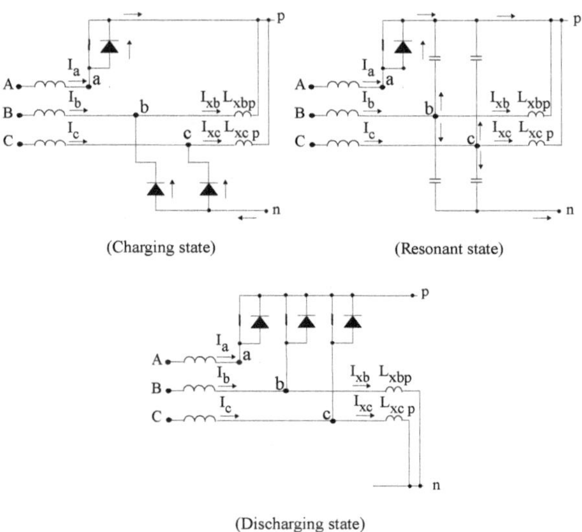

Fig. 3.17: The auxiliary circuit of ACS-QW-ARCP PWM converter during the commutation process.

$$t_4 = t_3 + \frac{Z_r\,|I_b| + V_{dc}}{\omega_r\,V_{dc}} \tag{3.78}$$

The duration of the discharging state is given by $\Delta t_{dis} = \dfrac{Z_r\,|I_{ph}| + V_{dc}}{\omega_r\,V_{dc}}$.

The device stresses described in section 3.1.3.2 by equations (3.73), (3.74) and (3.76) are also valid for the ACS-QW-ARCP circuit.

Since the operation of this converter does not principally differ from that described for the ACS-QW-ARC circuit, except that the phase current commutations are independent from each other, the analysis has no need to be verified by the simulation results. The operation waveforms of the converter during the analysed transition are similar to those shown in Fig. 2.21 for the inverter mode of operation.

3.2 Minimum Necessary Time Duration for ZVS Switching

The total duration of ZVS including charging, resonant and discharging state should not be longer than some percent of the switching period, for the duration of voltage transition affects the synthesis of the reference voltage and the quality of the generated AC voltage. The duration of the charging and discharging states also limits the minimum duration of the voltage vectors engaged before and after the resonant state. These are the subjects to be thoroughly discussed in chapter 5. In this section the circuits are inspected towards the total ZVS duration. For this purpose only the pole commutation concepts, i. e. the ACS-HW-ARCP and ACS-QW-ARCP circuits are considered. Other circuits are disregarded, because the ACS-QW-ARC circuit differs from its pole concept only in amplitude of the auxiliary switch current and the ACS-HW-ARC circuit is affected by switching of an undesired switching state.

For simplicity of discussion we assume that the circuits under consideration are ideal concluding I_{boost} and Δt are both zero. The circuits will be evaluated under *the same current peak for the auxiliary switches* which is assumed to be twice as large as the amplitude of phase current, i. e. $\mathbf{I_{S_x,max} = 2 \times \hat{I}_{ph}}$. So the auxiliary switches of both circuits have to withstand the same current stress.

Total duration of the zero voltage transition is calculated from the sum of the charging, resonant and discharging durations as follows:

$$\Delta t_{ZVS} = \Delta t_{ch} + \Delta t_{res} + \Delta t_{dis}. \tag{3.79}$$

a) The ARC-HW-ARCP Circuit:

The auxiliary circuit inductance can be calculated from (3.63) by substituting $I_{S_x,max} = 2 \times I_{ph}$ and $I_{boost} = 0$ in this equation. It results in:

$$L_x = \frac{1}{4}\frac{C_s V_{dc}^2}{I_{ph}^2}. \tag{3.80}$$

By substituting L_x in (3.45), (3.53) and (3.57), the duration of charging, resonant and discharging states can be found as follows:

$$\begin{cases} \Delta t_{ch} &= \dfrac{1}{2}\dfrac{C_s V_{dc}}{I_{ph}}, \\ \Delta t_{res} &= \pi\dfrac{C_s V_{dc}}{I_{ph}}, \\ \Delta t_{dis} &= \dfrac{1}{2}\dfrac{C_s V_{dc}}{I_{ph}}. \end{cases} \tag{3.81}$$

From there the total duration of the ZVS process is given by:

$$\Delta t_{HW,ZVS} = \frac{C_s V_{dc}}{I_{ph}}(1+\pi). \tag{3.82}$$

b) The ARC-QW-ARCP Circuit:

By substituting $I_{S_x,max} = 2 \times I_{ph}$ in (3.76) the value of L_x for this circuit is obtained as follows:

$$L_x = 2\frac{C_s V_{dc}^2}{I_{ph}^2}. \tag{3.83}$$

which is eight times larger than of auxiliary inductor of HW-concept. Substitution of L_x into (3.77), (3.70) and (3.78) yields the duration of charging, resonant and discharging states:

$$\begin{cases} \Delta t_{ch} &= 2\dfrac{C_s V_{dc}}{I_{ph}}, \\ \Delta t_{res} &= \pi\dfrac{C_s V_{dc}}{I_{ph}}, \\ \Delta t_{dis} &= 4\dfrac{C_s V_{dc}}{I_{ph}}. \end{cases} \tag{3.84}$$

and from there the total duration of the ZVS process in the ACS-QW-ARCP circuit is given by:

$$\Delta t_{QW,ZVS} = \frac{C_s V_{dc}}{I_{ph}}(6+\pi). \tag{3.85}$$

Comparison of equations (3.82) and (3.85) reveals that under the same current stress for the auxiliary switches the HW-concept is more than twice as fast as the QW-concept, if the boost energy and Δt are ignored.

3.3 Evaluation of the AC-Side Soft Commutated Circuits Concerning Switching Losses

In this section the ASSC PWM circuits are evaluated in an example from the point of view of losses. For this purpose, it is sufficient to consider only the switching and auxiliary circuit energy losses for one switching cycle, because the conduction energy losses of the main switches do not change so much with the auxiliary circuit used. Besides, the switching and auxiliary circuit losses are more concerned, because they dominantly limit the switching frequency in high frequency applications.

Since all the proposed circuits are switched at zero voltage switching conditions, it is assumed that all main switches are turned on under ZVS conditions. Hence, there are no switch turn-on and diode reverse-recovery energy losses in the main circuit of ASSC PWM converters. Gate losses of both main and auxiliary switches are small compared to switching and conduction losses and therefore will be neglected.

The circuits are evaluated in an example considering two commonly used semiconductor power devices, IGBT and Power MOSFET. In this example $V_{dc} = 800\,V$, $I_{a(max)} = 20\,A$, $I_b = -10\,A$, $I_c = -10\,A$ and the operation of converters in sector I of SVM voltage hexagon are assumed[3)] . Additionally, for a sufficiently small dV/dt and low turn-off loss of the main switches, an $8.2nF$ snubber capacitance C_S across the switches is assumed. For the main switches an IGBT (IRGPH50KD2, $1200V$ and $20A$) from International Rectifier with collector-to-emitter saturation voltage $V_{CE(on)} = 3.1V$ (at $T_j = 150^oC$ and $I_{CM} = 50A$) and antiparallel diode forward voltage drop $V_{FM} = 3.2V$ (at $T_j = 150^oC$) [47] is considered in case of IGBTs. In case of Power MOSFETs, a module from Siemens (BSM 191 (C), $1000V$ and $28A$) with an $R_{DS(on)} = 0.55\Omega$ (at $V_{GS} = 15V$ and $T_j = 100^oC$) and $C_{oss} = 0.8nF$ (at $V_{DS} = 35V$) is selected [48] for the main switches. The auxiliary switches will be appropriately selected for each topology in the following of this section.

[3)] For comparing the switching losses of the different ASSC PWM circuits with each other an example in medium power range which can be possibly applicated to all four ASSC PWM circuits is selected.

A maximum loading of $2.5 \times I_{a(max)}$ is assumed for the auxiliary circuits. This means that if the commutations of phase currents I_b and I_c need only one switching action in the auxiliary circuit (the case of non-pole commutated concepts), the current peak of the auxiliary switches is considered equal to $2.5 \times I_{a(max)} = 50A$. In case of commutated pole concepts, where these commutations need two switching actions in the auxiliary circuit, the current peak of the auxiliary switches is considered to be half of it, i.e. $1.25 \times I_{a(max)} = 25A$. In this way, the value of necessary silicon area, $A_{Silicon}$, remains the same for all auxiliary circuits when the same current density for the switches of different auxiliary circuits is assumed.

3.3.1 Device Energy Losses

3.3.1.1 IGBT

As mentioned, only the energy losses related to switching actions within one switching cycle are the concern of this section. From this point of view the conduction energy loss of the auxiliary switches, $E_{con,s}$, should be considered. This loss is obtained from the product of switch voltage drop times the current through the switch times conduction duration and given by:

$$E_{con,IGBT} = V_{CE(on)}\, I_{C(av)}\, \Delta t_{con,s}, \tag{3.86}$$

where $V_{CE(on)}$ and $I_{C(av)}$ are collector-to-emitter saturation voltage at the operation point given in device's data sheets and collector average current and Δt_{Con_s} is the conduction duration.

The turn-off switching energy loss is calculated parametrically, because the tail current, I_{tail}, and its duration cannot be derived from the device's data sheets. These parameters can be determined either experimentally or through the simulation results, when all necessary data for the simulation are known. The turn-off energy loss is calculated using linear fall and tail model of the IGBT's turn-off current. This model offers the adequate accuracy [44].

The expression of the turn-off energy loss, $E_{off,IGBT}$, is then given by (see also appendix A):

$$E_{off,IGBT} = \frac{I_{off}^2}{24\,C_S}\,\xi, \tag{3.87}$$

with:

$$\begin{cases} \xi &= t_{off}^2\,[(1-a)^2K(1-K) + a^2(1-K)(1+3K) + 6a(1-a)K(1-K)], \\ t_{off} &= t_t + t_{tail}, \\ a &= \dfrac{t_t}{t_{off}}, \\ K &= \dfrac{I_{tail}}{I_{off}}, \end{cases}$$

where I_{off}, t_{off} and C_S are turn-off current, turn-off duration and the snubber capacitance paralleled to the switch respectively.

The auxiliary diode conduction energy loss, $E_{con,d}$, is also calculated from the product of the diode average current, $I_{D(av)}$, times diode forward voltage drop, V_{FM}, times the conduction time $\Delta t_{con,d}$ as follows:

$$E_{con,d} = V_{FM}\,I_{D(av)}\,\Delta t_{con,d}. \tag{3.88}$$

3.3.1.2 Power MOSFET

The conduction energy loss of Power MOSFET is calculated using the rms-current ohm's law. Its turn-off energy loss, $E_{off,MOS}$, is also obtained from: (see appendix A)

$$E_{off,MOS} = \frac{I_D^2\,t_f^2}{24\,C_s}, \tag{3.89}$$

where I_D and t_f are drain current and current fall time and C_S is the snubber capacitance across the switches. Here also a linear fall turn-off current model is supposed.

If Power MOSFETs are used as auxiliary switches, their capacitive turn-on energy loss[4] can be calculated from:

$$E_{on,MOS} = \frac{1}{2} C_{oss} V_{DS}^2, \tag{3.90}$$

where C_{oss}[5] is the device output capacitance under the full drain-source voltage, V_{DS}.

3.3.2 Evaluation of the ARC-HW-ARC Circuit

As mentioned in section 2.3.1.2 the switching sequence $S_{1(pnn)} - S_{4(npp)} - S_{7(ppp)} - S_{2(ppn)} - S_{1(pnn)}$ should be used for switching this circuit in sector I of SVM voltage hexagon. The major losses appear during the transition $S_{4(pnn)} \rightarrow S_{1(npp)}$ which requires one switching action in the auxiliary circuit. It is also mentioned that in the switching state $S_{1(pnn)}$ the switches S_{ap}, S_{bn} and S_{cn} should remain on until the current I_{ax} reaches $I_{ch} = I_a + I_{boost}$. The current of the switches at the end of the charging state is then equal to the difference between the respective auxiliary and phase currents. Introducing $I_{S_x,max} = 50A$ in (3.40) and solving the equation system including (3.36), (3.35) and (3.40) results $L_x = 6.3\mu H$ and $I_{boost} = 12.6A$ for complete ZVS conditions. From there $Z_r = 19.6\Omega$ and $\omega_r = 3112697\, rad/s$ are obtained for the resonant circuit.

3.3.2.1 Main Circuit Switching Energy Losses

The main circuit switching energy losses can be listed as follows:

- Transition from $S_{1(pnn)}$ to $S_{4(npp)}$: turn-off energy loss of the main switches S_{ap}, S_{bn} and S_{cn} turned off under the surplus auxiliary currents at the end of boost charging,
- Transition from $S_{4(npp)}$ to $S_{7(ppp)}$: turn-off energy loss of the switch S_{an} turned off under I_a,

[4] In all considered circuits, the auxiliary switches are turned on under zero current condition, but the energy stored in their output capacitance, $E_{on,MOS}$, causes capacitive turn-on loss at turn-on. In case of Power MOSFETs, in converse to IGBTs, this energy cannot be ignored due to relatively large output capacitances.

[5] Note, that C_{oss} is dependent on V_{DS} during the turn-on transition, but as long as only the energy stored in the output capacitance under full drain-source voltage, and not the current produced by discharging this energy, is concerned, the value of C_{oss} under full V_{DS} can be used for calculating E_{on}.

- Transition from $S_{7(ppp)}$ to $S_{2(ppn)}$: turn-off energy loss of the switch S_{cp} turned off under I_c,
- Transition from $S_{2(ppn)}$ to $S_{1(pnn)}$: turn-off energy loss of the switch S_{bp} turned off under I_b.

Referring to (3.87) in case of IGBT, one can find the switching turn-off energy loss of the main switches, $E_{off(C1M,IGBT)}$, as follows:

$$E_{off(C1M,IGBT)} = \frac{1}{24 \cdot (8.2 \times 10^{-9})} [40\,\xi_b^{'} + 40\,\xi_c^{'} + 159\,\xi_a^{'} + 400\,\xi_a + 100\xi_b + 100\xi_c], \tag{3.91}$$

where $\xi_a^{'}$, $\xi_b^{'}$, $\xi_c^{'}$ correspond to the turn-off energy losses of the first transition and ξ_a, ξ_b, ξ_c correspond to the turn-off losses of the other transitions. They are defined separately because ξ according to (3.87) depends on the t_{off} which is different for different turn-off currents.

Since this circuit can be employed in all power ranges, both IGBT and Power MOSFET are considered for implementing the main switches. In case of Power MOSFETs the turn-off energy loss of the main switches, $E_{off(C1M,MOS)}$, can be calculated from (3.89) as follows:

$$E_{off(C1M,MOS)} = \frac{1}{24 \cdot (8.2 \times 10^{-9})} [159\,t_{fa}^{'2} + 40\,t_{fb}^{'2} + 40\,t_{fc}^{'2} + 400\,t_{fa}^2 + 100t_{fb}^2 + 100t_{fc}^2]. \tag{3.92}$$

3.3.2.2 Auxiliary Circuit Energy Losses

The ZVS duration, Δt_{ZVS}, can be good approximated by:

$$\Delta t_{ZVS} = \Delta t_{ch} + \Delta t_{dis} + \Delta t_{res} = 2.65\mu s, \tag{3.93}$$

in which Δt_{ch}, Δt_{dis} and Δt_{res} are the charging, discharging and the approximated resonant state durations respectively, using equation 3.94:

$$\begin{cases} \Delta t_{ch} = \Delta t_{dis} = \dfrac{3\,(I_a + I_{boost})\,L_x}{2\,V_{dc}} = 827\,ns, \\ \\ \Delta t_{res} = \pi\sqrt{2\,L_x\,C_s} = 1.0\mu s. \end{cases} \tag{3.94}$$

If a fast recovery rectifier diode (12FL60S02) [49] from International Rectifier is chosen for the auxiliary diodes, the sum of conduction energy losses of auxiliary diodes, $E_{con(C1A,D)}$, can be approximated by:

$$E_{con(C1A,D)} = \left(\frac{I_{S_x,max}}{2} + \frac{I_{S_x,max}}{4} + \frac{I_{S_x,max}}{4}\right) V_{FM}\,\Delta t_{ZVS} = 200\,\mu j, \tag{3.95}$$

where $I_{S_x,max} = 50\,A$ is the maximum current of the auxiliary switch and $V_{FM} = 1.5\,V$ is the forward voltage drop of the diodes. The auxiliary currents I_{xa}, I_{xb} and I_{xc} are assumed to vary linearly, so their average value during the commutation process will be half of their maximum.

The auxiliary switch is supposed to be implemented by a MOSFET Module BSM 191(C)[6)] [48] from Siemens. Using Power MOSFETs as auxiliary switch, the auxiliary circuit energy losses are the conduction energy loss of diodes and the capacitive turn-on, turn-off and conduction energy losses of the auxiliary switch.

The capacitive turn-on energy loss of the auxiliary switch, $E_{on(C1A,S_x,MOS)}$, can be estimated from the energy stored in its output capacitance, $C_{oss} = 0.8nF$, as follows:

$$E_{on(C1A,S_x,MOS)} = \frac{1}{2}\,C_{oss}\,V_{dc}^2 = 256\,\mu j. \tag{3.96}$$

Assuming the variation of the auxiliary currents linear, the conduction energy loss of the switch, $E_{con(C1A,S_x,MOS)}$, is given by:

$$E_{con(C1A,S_x,MOS)} \;=\; 2\int_0^{\frac{\Delta t_{ZVS}}{2}} R_{DS(on)} \left(\frac{2\,I_{S_x,max}t}{\Delta t_{ZVS}}\right)^2 \tag{3.97}$$

6) Note, that the repetitive rating of pulsed drain (collector) current of devices is 3-4 times higher than their continuous drain (collector) current. Because the duration of commutation process is very short compared to switching cycle, the selected device can easily withstand $I_{S_x,max}$.

$$= \left(\frac{I^2_{S_x,max}}{3}\right) R_{DS(on)}\,\Delta t_{ZVS} = 817\,\mu j.$$

Since the switch turn-off current at the end of discharging state is very small and its current fall time is relative short, the turn-off energy loss can be ignored.

Thus the total auxiliary circuit energy losses for one switching cycle, $E_{ac(C1A,MOS)}$, can be given by:

$$E_{ac(C1A,MOS)} = E_{con(C1A,D)} + E_{con(C1A,S_x,MOS)} + E_{on(C1A,S_x,MOS)} = 1.27\,mj. \tag{3.98}$$

In case the auxiliary switch is implemented with an IGBT, for example the IGBT (IRGPH40F)[6] [50] from International Rectifier with $V_{CE(on)} = 3.2\,V$, the conduction energy loss of the switch, $E_{con(C1A,S_x,IGBT)}$, is calculated as follows:

$$E_{con(C1A,S_x,IGBT)} = \frac{I_{S_x,max}}{2} V_{CE(on)}\,\Delta t_{ZVS} = 205\,\mu j. \tag{3.99}$$

If it is assumed that $I_{S_x,off} = 0.1 \times I_{S_x,max} = 5\,A$, the turn-off energy loss of the switch, $E_{off(C1A,S_x,IGBT)}$, can be estimated from the given waveforms in [50] (considering 60^oC case temperature):

$$E_{off(C1A,S_x,IGBT)} = 950\,\mu j. \tag{3.100}$$

The total auxiliary circuit energy losses in case of IGBTs, $E_{ac(C1A,IGBT)}$, are given by:

$$E_{ac(C1A,IGBT)} = E_{con(C1A,D)} + E_{con(C1A,S_x,IGBT)} + E_{off(C1A,S_x,IGBT)} = 1.97\,mj. \tag{3.101}$$

3.3.3 Evaluation of the ACS-HW-ARCP Circuit

Considering the same switching sequence $S_{1(pnn)} - S_{7(ppp)} - S_{2(ppn)} - S_{1(pnn)}$ as used for the analysis of this circuit in section 3.1.2, the energy losses produced during the transition between $S_{1(pnn)}$ and $S_{7(ppp)}$ are the major energy losses of the circuit. This is because of the actuation of the auxiliary circuit which is accompanied with two switching actions.

Because of six switches in the auxiliary circuit, this converter is commonly employed in high power applications and IGBT is used as main switches. For this reason, the case of Power MOSFET as main switch is not considered for this converter.

The auxiliary switches of this converter are turned on and off under ZCS condition and the voltage on the auxiliary switches is now $V_{dc}/2$, thus the IGBT (IRGBC30UD2)[6)] [51] from International Rectifier can be selected for the implementation of the auxiliary switches. The collector-to-emitter saturation voltage $V_{CE(on)}$ and the Forward voltage drop of the antiparallel diode V_{FM} of this IGBT are $3\,V$ and $1.7\,V$ respectively.

After substituting $|I_{b,c}| = 10\,A$, $I_{S_x,max} = 25\,A$ and $\Delta t = 100\,ns$ in (3.59), (3.60) and (3.63) and solving them with (3.53), this results in $L_x = 6.5\,\mu H$, $I_{boost} = 9.5\,A$ and $\Delta t_{res} = 1.15\,\mu s$.

3.3.3.1 Main Circuit Switching Energy Losses

The main circuit switching energy losses are listed as follows:

- Transition from $S_{1(pnn)}$ to $S_{7(ppp)}$: turn-off energy loss of the main switches S_{bn} and S_{cn} turned off under the surplus auxiliary currents at the end of boost charging,
- Transition from $S_{7(ppp)}$ to $S_{2(ppn)}$: turn-off energy loss of the main switches S_{cn} turned off under I_c,
- Transition from $S_{2(ppn)}$ to $S_{1(pnn)}$: turn-off energy loss of the main switches S_{bn} turned off under I_b.

Analogous to the first topology discussed in the last section, the main circuit switching energy loss for the use of IGBTs, $E_{off(C2M,IGBT)}$, is given as follows:

$$E_{off(C2M,IGBT)} = \frac{1}{24 \cdot (8.2 \times 10^{-9})} [33.6\,\xi_b' + 33.6\,\xi_c' + 100\,\xi_b + 100\,\xi_c], \quad (3.102)$$

3.3.3.2 Auxiliary Circuit Energy Losses

The total duration of transition Δt_{ZVS} is calculated from:

$$\begin{cases} \Delta t_{ZVS} &= \Delta t_{ch} + \Delta t_{dis} + \Delta t_{res} = 1.95\mu s, \\ \Delta t_{ch} &= \dfrac{2\left(|I_b| + |I_c| + I_{boost}\right) L_x}{V_{dc}} = 480\, ns, \\ \Delta t_{dis} &= \dfrac{2\left(|I_b| + |I_c|\right) L_x}{V_{dc}} = 325\, ns, \\ \Delta t_{res} &= 1.15\mu s \text{ resulted from (3.53)}. \end{cases} \tag{3.103}$$

The conduction energy loss of the auxiliary switches, $E_{con(C2A,S_x)}$, and of the antiparallel diodes, $E_{con(C2A,D)}$, can be given as follows:

$$E_{con(C2A,S_x)} = 2\left(\frac{I_{S_x,max}}{2} V_{CE(on)}\, \Delta t_{ZVS}\right) = 147\, \mu j, \tag{3.104}$$

$$E_{con(C2A,D)} = 2\left(\frac{I_{S_x,max}}{2} V_{FM}\, \Delta t_{ZVS}\right) = 83\, \mu j. \tag{3.105}$$

Then the total auxiliary circuit energy losses, $E_{ac(C2A)}$, is given by:

$$E_{ac(C2A)} = E_{con(C2A,D)} + E_{con(C2A,S_x)} = 230\, \mu j. \tag{3.106}$$

Since the auxiliary switches of this circuit are implemented by IGBTs, the capacitive turn-on energy loss of the auxiliary switches is neglected.

3.3.4 Evaluation of the ACS-QW-ARC Circuit

Considering the same switching sequence $S_{1(pnn)} - S_{7(ppp)} - S_{2(ppn)} - S_{1(pnn)}$, in the transition $S_{1(pnn)} \rightarrow S_{7(ppp)}$, the switches S_{bn} and S_{cn} can be turned off under ZCS conditions before their antiparallel diode turns off resulting in less switch turn-off actions and consequently less switch turn-off energy loss.

Since the converter has no need for an additional voltage source at the auxiliary circuit in the rectifier mode operation for recovering the commutation energy, and in order to realize the same conditions for all circuits during the discharging state, the sources $P1$ and $P2$ in Fig. 3.12 will not be considered in the discussion of this section.

As aforementioned in chapter 2, due to hard turn-off conditions at the auxiliary circuit of this converter, the auxiliary switches have to be implemented with Power MOSFETs. This limits the application of this converter to medium power range. In this power range both IGBT and Power MOSFET can be considered as main switches.

3.3.4.1 Main Circuit Switching Energy Losses

The main circuit switching energy losses can be listed now as follows:

- Transition from $S_{1(pnn)}$ to $S_{7(ppp)}$: no loss,
- Transition from $S_{7(ppp)}$ to $S_{2(ppn)}$: turn-off energy loss of the main switch S_{cp} turned off under I_c,
- Transition from $S_{2(ppn)}$ to $S_{1(pnn)}$: turn-off energy loss of the main switch S_{bp} turned off under I_b.

Referring to (3.87), the main switch turn-off energy losses in case of IGBTs, $E_{off(C3A,IGBT)}$, and Power MOSFETs, $E_{off(C3A,MOS)}$, are given by following expressions:

$$E_{off(C3A,IGBT)} = \frac{1}{24 \cdot (8.2 \times 10^{-9})}[100\,\xi_b + 100\,\xi_c], \tag{3.107}$$

$$E_{off(C3A,MOS)} = \frac{1}{24 \cdot (8.2 \times 10^{-9})}[100\,t_{fb}^2 + 100\,t_{fc}^2]. \tag{3.108}$$

3.3.4.2 Auxiliary Circuit Energy Losses

Considering the following equation giving the ZVS duration:

$$\begin{cases} \Delta t_{ZVS} &= \Delta t_{ch} + \Delta t_{dis} + \Delta t_{res}, \\ \Delta t_{ch} &= \dfrac{I_{b,c}\, L_x}{V_{dc}}, \\ \Delta t_{dis} &= \dfrac{I_{S_x,max}\, L_x}{2\, V_{dc}}, \\ \Delta t_{res} &= \dfrac{\pi}{2}\, \sqrt{2\, L_x\, C_s}, \end{cases} \tag{3.109}$$

and solving the equation system including these equations and (3.75), while substituting $I_{S_x,max} = 50\, A$, results in $L = 46\, \mu H$, $\Delta t_{ch} = 575\, ns$, $\Delta t_{res} = 1364\, ns$, $\Delta t_{dis} = 1437\, ns$ and $\Delta t_{ZVS} = 3.37\, \mu s$.

Considering the same diodes and Power MOSFET as in section 3.3.1.2 for the auxiliary diodes and switches and performing the same calculations, it is found:

$$E_{con(C3A,D)} = 2\, \frac{I_{S_x,max}}{4}\, V_{FM}\, \Delta t_{ZVS} + \frac{I_{S_x,max}}{2}\, V_{FM}\, \Delta t_{dis} = 180\, \mu j, \tag{3.110}$$

$$E_{on(C3A,S_x)} = \frac{1}{2}\, C'_{oss}\, V_{dc}^2 = 256\, \mu j, \tag{3.111}$$

$$\begin{aligned} E_{con(C3A,S_x)} &= \int_0^{\Delta t_{ch}+\Delta t_{res}} R_{DS(on)} \left(\frac{I_{S_x,max} t}{\Delta t_{ch} + \Delta t_{res}} \right)^2 \\ &= \frac{I_{S_x,max}^2}{3}\, R_{DS(on)}\, (\Delta t_{ch} + \Delta t_{res}) = 1.27\, mj. \end{aligned} \tag{3.112}$$

The turn-off energy loss of the auxiliary switch is assumed to be negligible. The total auxiliary circuit energy losses, $E_{ac(C3A)}$, are given by:

$$E_{ac(C3A)} = E_{con(C3A,D)} + E_{on/con(C3A,S_x)} = 1.74\, mj. \tag{3.113}$$

3.3.5 Evaluation of the ARC-QW-ARCP Circuit

The same switching sequence $S_{1(pnn)} - S_{7(ppp)} - S_{2(ppn)} - S_{1(pnn)}$ is considered. It must be noted that one may consider the sequence $S_{1(pnn)} - S_{2(ppn)} - S_{7(ppp)} - S_{1(pnn)}$ in which the soft current commutation at phases b and c takes place separately in the first and second transitions. This results in the same losses, because the commutation of phase currents occurs independently. For this reason the losses resulting from the commutation of phase currents with the same magnitude are also equal.

Like the ARC-QW-ARC circuit this circuit is also applicable to medium power range. Therefore, for the main switches both the cases IGBT and Power MOSFET are considered. The auxiliary switches have to be implemented with Power MOSFET due to the hard turn-off conditions.

3.3.5.1 Main Circuit Switching Energy Losses

The energy losses listing remains as it was given for the ACS-QW-ARC circuit. The turn-off energy losses of the main switches for IGBTs and Power MOSFETs are also the same as those given by (3.107) and (3.108).

3.3.5.2 Auxiliary Circuit Energy Losses

Introducing $I_{S_x,max} = 25\,A$ in (3.76) results in $L_x = 46\,\mu H$. The durations of ZVS stages are then given as follows:

$$\begin{cases} \Delta t_{ZVS} &= \Delta t_{ch} + \Delta t_{dis} + \Delta t_{res} = 3.37\,\mu s, \\ \Delta t_{ch} &= \dfrac{I_{b,c}\,L_x}{V_{dc}} = 575\,ns, \\ \Delta t_{dis} &= \dfrac{I_{S_x,max}\,L_x}{V_{dc}} = 1437\,ns, \\ \Delta t_{res} &= \dfrac{\pi}{2}\sqrt{2\,L_x\,C_s} = 1364\,ns. \end{cases} \tag{3.114}$$

Using the same Power MOSFETs and diodes as used in section 3.3.1.2 and considering the design considerations mentioned in section 2.3.4.2 the energy losses of the auxiliary circuit resulting from the commutation of both phase currents I_b and I_c are given as follows:

$$E_{con(C4A,D)} = 2\,V_{FM}\left(\frac{I_{S_x,max}}{2}\,(\Delta t_{ZVS} + 2\,\Delta t_{dis})\right) = 265\,\mu j, \tag{3.115}$$

$$E_{on(C4A,S_x)} = 2\,\frac{1}{2}\,C_{oss}^{'}\,V_{dc}^2 = 512\,\mu j, \tag{3.116}$$

$$E_{con(C4A,S_x)} = 2\left(\frac{I^2_{S_x,max}}{3}\right) R_{DS(on)}\left(\Delta t_{ch} + \Delta t_{res}\right) = 299\,\mu j. \qquad (3.117)$$

And the total auxiliary circuit energy losses, $E_{ac(C4A)}$, are calculated by:

$$E_{ac(C4A)} = E_{con(C4A,D)} + E_{on/con/off(C4A,S_x)} = 1.076\,mj. \qquad (3.118)$$

3.3.6 Comparison

To give a simple numerical comparison among the switching losses resulting from the different ASSC circuits within one switching cycle, the following assumptions are made:

1. dI/dt of I_{fall} and I_{tail} of the selected IGBTs dictated by the gate circuit remains constant for all turn-off currents,
2. In Equation (3.87) : $t_{off} = t_{off}\Big|_{I_C=20A} (= 660ns) \times \frac{I_C}{20A}$ (proportional), $a = 0.3$ and $K = 0.25$.
3. For the use of Power MOSFET: $t_f = t_f\Big|_{I_D=20A} (= 60ns) \times \frac{I_D}{20A}$ (proportional).

The obtained results are summarized in the table 3.1. The elements without any result in this table are the unsuitable cases, which are not considered in the discussion of this section. As can be seen from the table, under the same loading of the auxiliary circuits, the ACS-HW-ARCP circuit has the least switching losses in the auxiliary circuit due to half DC voltage on the auxiliary switches and ZCS switching conditions in the auxiliary circuit. The ZCS conditions in the auxiliary circuit makes implementation of the auxiliary switches with IGBTs possible, so the circuit enjoys the low turn-on and conduction losses of these devices in high power applications. This is achieved however at the expense of more turn-off actions of main switches and complexity of control timing in this circuit. Because of lack of the ZCS conditions in the auxiliary circuit of the ACS-QW-ARC/-ARCP circuits, the auxiliary switches have to be implemented with power MOSFETs. The high turn-on and conduction losses of power MOSFETs increase then the total energy losses of the auxiliary circuit considerably.

Comparing the pole commutated circuits (ACS-HW/QW-ARCP) being almost complex to the same extent, one can simply conclude that the lower total

Comparison of ASSC Circuits under the Same A_S [7]					
Topology ACS-		HW-ARC	HW-ARCP	QW-ARC	QW-ARCP
Circuit Characteristics					
L_x		$6.3\,\mu H$	$6.5\,\mu H$	$46\,\mu H$	$46\,\mu H$
$I_{S_x,max}$		$50\,A$ [6]	$25\,A$	$50\,A$ [6]	$25\,A$
Δt_{ZVS}		$2.65\,\mu s$	$1.95\,\mu s$	$3.37\,\mu s$	$3.37\,\mu s$
Switching Energy Losses for One Switching Cycle					
Main	*IGBT*	$458\,\mu j$	$53\,\mu j$	$44\,\mu j$	$44\,\mu j$
Circuit	*MOSFET*	$7.6\,\mu j$	-	$0.7\,\mu j$	$0.7\,\mu j$
Auxiliary	*IGBT*	$1.97\,mj$	$0.23\,mj$	-	-
Circuit	*MOSFET*	$1.27\,mj$	-	$1.74\,mj$	$1.07\,mj$

Table 3.1: Comparison of ASSC PWM circuits under the same loading of the auxiliary circuit.

switching losses of ACS-HW-ARCP circuit in table 3.1 shows in fact the ability of the circuit for operation at higher switching frequencies considering the same peak current and cooling system for the auxiliary switches of both circuits. Since the waveform quality of the produced AC voltages and currents raises with the increase of the switching frequency, this represents a great advantage.

Another important point is the duration of ZVS commutation, Δt_{ZVS}. As will be shown in chapter 5, the space vector modulation is directly affected by the duration of resonant commutations. A shorter ZVS duration is therefore desirable. This is another advantage of the ACS-HW-ARCP circuit under this circumstance. It is also worth noting that Δt_{ZVS} can be certainly reduced at the expense of increasing the current peak in the auxiliary circuit by selecting smaller auxiliary inductors, but the ACS-HW-ARCP circuit has under the same peak current in the auxiliary inductors the shortest duration for the ZVS process.

A comparison among the ASSC circuits with the same auxiliary inductors can also be interesting from the point of view of the maximum current and voltage stresses in the main circuit. This comparison is also carried out by assuming the

[7] Effectively used silicon area in the auxiliary circuit.

Comparison of ASSC Circuits under the Same Auxiliary Inductors					
Topology ACS-		HW-ARC	HW-ARCP	QW-ARC	QW-ARCP
Circuit Characteristics					
$I_{S_x,max}$		$45\,A^{6)}$	$22\,A$	$85\,A^{6)}$	$42\,A^{6)}$
Δt_{ZVS}		$2.5\,\mu s$	$2.3\,\mu s$	$1.3\,\mu s$	$1.3\,\mu s$
$(dI/dt)_{max}$		$53\,A/\mu s$	$40\,A/\mu s$	$80\,A/\mu s$	$80\,A/\mu s$
$(dV/dt)_{max}$		$381\,V/\mu s$	$773\,V/\mu s$	$2\,kV/\mu s$	$2\,kV/\mu s$
Switching Energy Losses for One Switching Cycle					
Auxiliary	*IGBT*	$1.13\,mj$	$0.24\,mj$	-	-
Circuit	*MOSFET*	$1.03\,mj$	-	$1.07\,mj$	$0.99\,mj$

Table 3.2: Comparison of ASSC PWM circuits with the same auxiliary inductors ($L_x = 10\,\mu H$).

auxiliary inductors of all circuits equal to $10\,\mu H$ in the same example. The characteristics of the resonant circuit, $Z_r = 24.7\,\Omega$ and $\omega_r = 2469324\,rad/s$, remain the same for all circuits. The results are shown in table 3.2.

With the same auxiliary inductors the ACS-QW-ARCP offers a shorter duration for ZVS commutation at the expense of increased dV/dt and dI/dt for the main switches and of higher current peak in the auxiliary switches compared to ACS-HW-ARCP circuit. Even under the same auxiliary inductors the auxiliary circuit energy losses of the ACS-HW-ARCP circuit is less than that of other circuits, because the auxiliary switches of this circuit are turned off under complete ZCS conditions. This is not the case of the other circuits under study. The switching losses of the main circuits do not differ so much from those given in table 3.1.

The results given in this section reveal that the ACS-HW-ARCP converter is the best choice among the ASSC PWM converters despite six active switches at the auxiliary circuit. Previous research works have also shown the superiority of this circuit as compared to the DC-side soft commutated converters [52] and [53].

3.4 Summary of Chapter

This chapter gives a detailed analysis for the four AC-Side Soft-Commutated (ASSC) PWM Converters described in last chapter. While the circuit design equations are explored, the voltage and current stresses on the auxiliary switches are also given for each topology. The expressions which determine the necessary value of boost energy depending on the DC link voltage, load current and the auxiliary circuit resonant characteristics are derived for the Half-Wave circuits.

The four AC-Side Soft-Commutated (ASSC) PWM Converters are evaluated regarding the switching losses. The results of the switching loss calculation show that the AC-Side Half-Wave Auxiliary Resonant Commutated Pole (ACS-HW-ARCP) converter has the advantages of the lowest total switching losses including the losses of auxiliary circuit and the lowest voltage and current stresses on the auxiliary switches. For this reason this converter can be the best candidate for high power applications. The AC-Side Quarter-Wave Auxiliary Resonant Commutated Pole (ACS-QW-ARCP) converter can also be an alternative in medium power range applications. Both circuits are commutated pole concept and therefore need six switches in the auxiliary circuit.

Lower switching losses in the AC-Side Half-Wave Auxiliary Resonant Commutated Pole (ACS-HW-ARCP) converter shows the ability of this circuit for operating in a higher switching frequency.

From the point of view of ZVT duration, the Quarter Wave (QW)-concept is faster than the Half-Wave (HW) concept under the same resonant circuit characteristics (Z_r and ω_r). But this advantage is achieved at the expense of a almost two fold higher current stress on the auxiliary switches of Quarter Wave (QW)-concept. Considering the same current stress for the auxiliary switches of both concepts, the Half Wave (HW)-concept will be faster than Quarter Wave (QW)-concept.

4. Comparative Study of SVM Techniques

In the previous chapters the ASSC-PWM circuits were introduced and well analysed. The advantages and weak points of each circuit were also enumerated. Following the goal of this work, namely the investigation and modification of the ASSC-PWM converters to find the best one from different points of view, it was shown in the last chapter that the ASC-HW-ARCP represents the best choice in high power applications from the losses point of view. The ASC-QW-ARCP can be also an alternative in the medium power range. One of the mentioned topological advantages of the both circuits is their ability to operate with every switching technique. In other words, their auxiliary circuit makes no condition for the switching method used. Now this question arises which switching technique represents the highest performance in terms of circuit losses and waveforms quality.

This chapter deals with the answer of this question and presents a detailed investigation of the important Space Vector PWM (SVPWM) techniques. It starts with a short background and discusses the space vector modulation as well as degrees of freedom of modulation used by different techniques for optimizing the modulation strategy. It is followed by an introduction of Continuous and Discontinuous SVPWM (C-SVPWM and D-SVPWM) techniques utilizing the degrees of freedom of modulation in different ways. The D-SVPWM techniques will be thoroughly discussed and compared with each other and with continuous SVPWM as well. This chapter also shows the impacts of a Minimum Loss D-SVPWM technique on reducing the switching losses in main and auxiliary circuits of the converters.

Since the discussion of modulation techniques is commonly based on the inverter operation of AC/DC converters, the discussion of this chapter is also based on the PWM voltage source inverter shown in Fig. 4.1 in which the directions

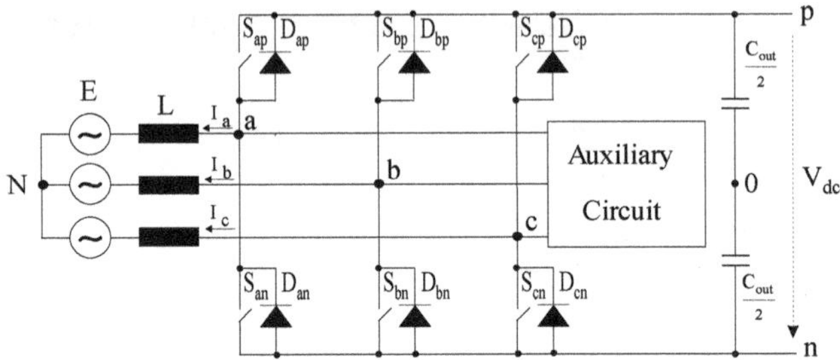

Fig. 4.1: Three-Phase Inverter Circuit.

of the phase currents towards the load are assumed positive. The letters p and n corresponding to polarity of phase voltages are also omitted from the index of voltage vectors for simplicity.

In order to make the results of this chapter more expressive, all AC quantities and duration of voltage vector shown in the figures are normalized to the respective base values described in appendix B.

4.1 Background

In a three phase soft switching PWM voltage source inverter shown in Fig. 4.1 the phase voltages at nodes a, b and c are generated by switching the bridge legs between the positive and negative DC bus rails according to a switching pattern, so that the locally averaged phase voltages $\bar{V}_a$, $\bar{V}_b$ and $\bar{V}_c$ correspond to the desired reference values which are usually three symmetrical, sinusoidal voltages with variable amplitude and/or frequency.

The switching pattern can be generated through the different modulation techniques, but all these techniques follow the common aims which are listed as follows:

- The desired load voltage should be generated with as little switching actions as possible to minimize the switching losses in the inverter.

- For a given DC bus voltage the maximum amplitude of the fundamental component of the phase voltage should be as high as possible to fully utilize the DC bus voltage and to make the highest maximum linear modulation index, m_{max}, available.

- The current ripples should be kept as low as possible, because they produce additional losses in the load.

- Low frequency harmonics should be avoided because they can cause torque pulsations in a motor load.

- The modulation principles should be easy for implementation in a microprocessor and should consume little computation time.

Also, conformity to the auxiliary circuit requirements and minimum number of switching action in the auxiliary circuit are of the significant aims followed by every high performance PWM technique used in soft commutated inverters.

The PWM techniques employ the "per sampling cycle volt-second balance" principle to generate the desirable output voltage waveform. In the triangular intersection technique known as Sinusoidal PWM (SPWM), the reference voltage waveforms are compared with the triangular carrier wave and the intersections define the switching instants. In this technique, the maximum value of the amplitude of the fundamental component of the inverter phase voltage, $\hat{V}_{a(1),max}$, has to be restricted to $0.5\,V_{dc}$ (when the modulation index m=1) in order to obtain a sinusoidal output voltage waveform. This value is only 78.5% of the maximum value of the fundamental component amplitude of the six-step mode voltage, $\frac{2}{\pi}V_{dc}$ (see Fig. 4.2(a and b)). The SPWM technique has been well investigated in published works such as [56, 57, 58].

The star connected loads without the neutral wire provide a degree of freedom for all PWM techniques in determining the duty cycle of the inverter switches. In the triangular intersections method, this degree of freedom appears in choosing the reference voltage wave. Adding a zero sequence signal[1)] to the three phase reference signals does not change the inverter output line-to-line voltage, but influences the modulation linearity range, waveform quality (current ripples) and switching losses. Utilizing this degree of freedom, the Harmonic Injected SPWM

[1)] A signal with the frequency $3 \times f_1$ (f_1: fundamental frequency, see also appendix D).

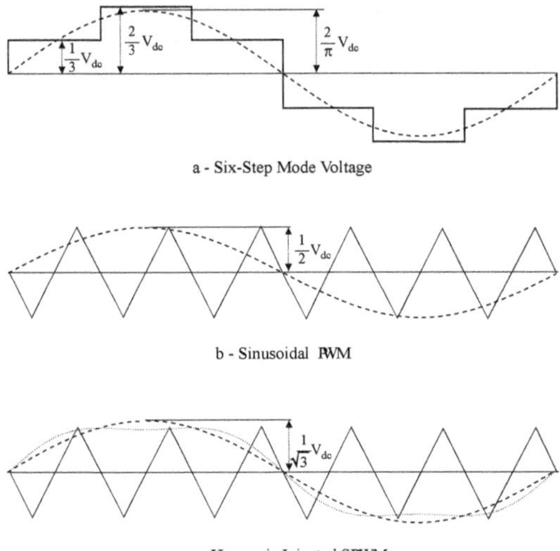

Fig. 4.2: Maximum value of the fundamental component amplitude of the phase voltage $\hat{V}_{a(1),max}$ *for a) the six-step mode voltage, b) linear range of SPWM technique and c) linear range of HI-SPWM technique.*

(HI-SPWM) technique was employed in [75] by mixing triple harmonics with the modulating waveforms in order to increase the output voltage in the linear range. Taking the third harmonic for example, the magnitude and phase of the injected third harmonic are determined in such a way that the highest value for the amplitude of the fundamental component of output voltage can be attained. [59] addresses the new reference voltage expression for this example as follows:

$$V_{ref}(t) = \hat{V}_{ref}\left(\sin(\omega_1 t) + \frac{1}{6}\sin(3\,\omega_1 t)\right). \qquad (4.1)$$

This method is linear up to 90.7% of a six-step mode voltage, and therefore, is superior to SPWM, which is only linear up to 78.5% (see Fig. 4.2(c)).

The development of microprocessors facilitates the direct digital technique known as Space Vector PWM (SVPWM) by employing the complex voltage vector for the "per sampling cycle volt-second balance" principle discussed in

[60, 61, 62]. In this technique, the concept of complex plane voltage vectors is utilized to calculate the duty cycle of the inverter switching devices and digital counters utilize the duty cycle informations to program the switch gate signals.

The advantages of SVPWM including dismissal of 2/3 phase transformation required in a field oriented control (FOC) system and extension of the linear modulation range up to 90.07% of a six-step mode voltage, have made this technique very popular in the past years.

4.2 The Basic Principle of Space Vector PWM

Every symmetrical three phase reference voltage with the amplitude $|V_{ref}|$ and the frequency ω_1 produces a rotating voltage vector $\vec{V}_{ref}$ in the $\alpha\beta$-plane with the length of $|V_{ref}|$ and angle $\phi = \omega_1 t$.

The principle of the SVPWM is to generate the reference voltage vector $\vec{V}_{ref}$ with the eight voltage vectors $\vec{V}_0 ... \vec{V}_7$ shown in Fig. 2.3. For this purpose the reference vector $\vec{V}_{ref}$ is sampled in discrete time steps, namely sampling cycle T_S. The sampled vector in the nth sample interval, $\vec{V}_{ref}(n\,T_S)$, is generated using its adjacent voltage vectors. When it is located in sector I, for example, it is generated by the vectors $\vec{V}_1$ and $\vec{V}_2$ as shown in Fig. 4.3.

In every sample interval the required average values of the length of the vectors $\vec{V}_1$ and $\vec{V}_2$ are set by varying their duty cycles T_1 and T_2 according to the principle of volt-second balance as follows:

$$T_S\,\vec{V}_{ref} = T_1\,\vec{V}_1 + T_2\,\vec{V}_2. \tag{4.2}$$

It is worth mentioning that theoretically the reference voltage vector $\vec{V}_{ref}$ can be generated by the voltage vectors $\vec{V}_0 \cdots \vec{V}_7$ in a great number of ways, but there are two reasons for generating $\vec{V}_{ref}$ using its adjacent voltage vectors:

- First, since the sum of duty cycles of the voltage vectors is limited to T_S, it results in the smallest absolute value for the sum of the duty cycles (also the lengths) of the vectors generating $\vec{V}_{ref}$. This maximizes the largest magnitude of $\vec{V}_{ref}$ which can be generated by the inverter and hence, makes the highest linear modulation index available.

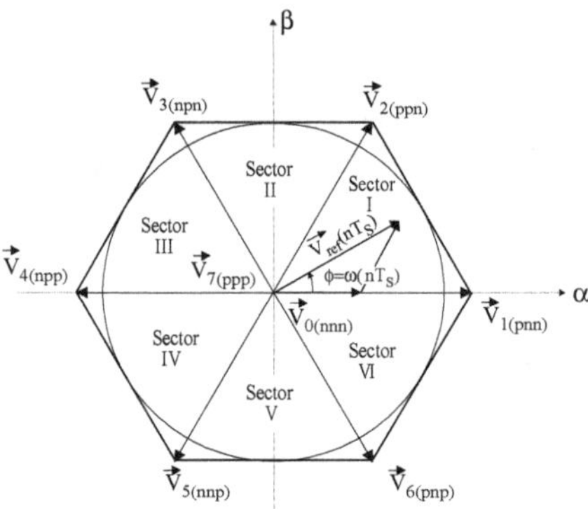

Fig. 4.3: Generation of the reference voltage vector $\vec{V}_{ref}$.

- Second, it has been proven in [63] that the best quality for AC voltages can be thereby achieved.

As long as $\vec{V}_{ref}(n\,T_S)$ is located inside the voltage hexagon, the sum of duty cycles T_1 and T_2 is smaller than T_S. For the remaining time of the sample interval, the zero vectors $\vec{V}_0$ and/or $\vec{V}_7$ have to be switched.

In a general form, the duty cycles T_K, T_{K+1} and $T_{0/7}$, belonging to the adjacent vectors of $\vec{V}_{ref}$, $\vec{V}_K$ and $\vec{V}_{K+1}$, and to the zero vector $\vec{V}_{0/7}$ respectively, are given at any instant by:

$$\left\{\begin{array}{rcl} T_K & = & \frac{\sqrt{3}}{2}\,T_S\,m\,\sin\left(\frac{K\,\pi}{3}-\phi\right), \\ T_{K+1} & = & \frac{\sqrt{3}}{2}\,T_S\,m\,\sin\left(\phi-\frac{(K-1)\,\pi}{3}\right), \\ T_{0/7} & = & T_7+T_0 \;=\; T_S-T_K-T_{K+1}, \end{array}\right. \tag{4.3}$$

where K is the number of the sector in which $\vec{V}_{ref}$ is located (if $K=6$ than $K+1=1$) and

$$m = \frac{|\vec{V}_{ref}|}{\frac{V_{dc}}{2}}, \tag{4.4}$$

is the modulation index.

With the restriction $T_K + T_{K+1} \leq T_S$ for any angle ϕ the maximum length of the vector $\vec{V}_{ref}$ is confined to $V_{dc}/\sqrt{3}$ which corresponds to the maximum linear modulation index $m_{max} = 1.15$.

The first degree of freedom of modulation, i. e. generation of $\vec{V}_{ref}$ by an optional combination of the voltage vectors, has been utilized to maximize the available linear modulation index by selecting the adjacent voltage vectors of $\vec{V}_{ref}$. Now, it can be seen that three further degrees of freedom are still available for optimizing the modulation strategy:

- The selected voltage vectors can be applied in different sequences.
- For the duration, $T_{0/7}$, one of the two voltage vectors $\vec{V}_0$ and $\vec{V}_7$ or a combination of both can be switched.
- The length of the sampling interval T_S can be varied in several ways.

One of the aforementioned aims of PWM techniques was generation of AC voltages at the desired quality while keeping the switching frequency as low as possible in order to minimize the switching losses. For this purpose, the number of switching actions (the number of bridge legs having to be switched) between different inverter switching states should be kept as low as possible, thereby the total losses including the auxiliary circuit losses are minimized. An optimal sequence, which needs only one switching action between every two different switching states, is applicable only to the commutated pole concepts of ASSC PWM circuits (ACS-HW/QW-ARCP). For others, the switching sequence has to be modified in such a way that turn-on actions of some switches become synchronized, thereby the modulation technique can meet the auxiliary circuit requirements. The switching sequences described in sections 2.3.1.2 and 2.3.3.2 of chapter 2 for the ACS-HW-ARC and ACS-QW-ARC circuits need more than one switching action between some of the switching states because they utilize the second and third degree of freedom of the modulation to meet the auxiliary circuit requirements. As a result, the second and third degree of freedom of modulation can be utilized either for optimizing the modulation strategy or for creating a modified SVPWM technique which facilitates a certain auxiliary circuit at all.

As stated before, the absence of neutral wire in star connected loads provides a degree of freedom in determining the duty cycles of the inverter switches. In the space vector PWM this degree of freedom appears by choosing and portioning of the zero voltage vectors $\vec{V}_0$ and $\vec{V}_7$. The choice of the zero voltage vector affects both the number of switching actions within the sampling cycle and the local average value of the common mode voltage V_{N0} (see appendix D) according to the following description for the symmetrical loads:

$$V_{N0} = \frac{V_{dc}}{2}\left(t_7 + \frac{(-1)^K}{3} t_K + \frac{(-1)^{K+1}}{3} t_{K+1} - t_0\right), \tag{4.5}$$

where t_K and t_{K+1} are the normalized duty cycles of non-zero voltage vectors $\vec{V}_K$ and $\vec{V}_{K+1}$ and t_7 and t_0 are the normalized duty cycles of zero vectors $\vec{V}_7$ and $\vec{V}_0$ respectively. The common mode voltage V_{N0} also influences the spectrum of the output voltage. The choice of the zero voltage vector leading to different SVPWM technique will be discussed in detail in the next section.

The fourth degree of freedom of the modulation allows the power spectrum of the output voltages to be improved by varying the length of the sampling interval. Variation of sampling interval, which can be either symmetrical or dependent on the modulation index or even random, scatters the first maximum of high order harmonics around the sampling frequency and smoothes the spectrum of AC quantities. This issue is out of framework and patience of this work. For further discussion in this regard see [64, 65].

4.3 Introduction of Space Vector Modulation Techniques

4.3.1 Continuous Modulation

The continuous Modulation technique is the conventional SVPWM, in which the switching sequence: $S_0 \rightarrow S_1 \rightarrow S_2 \rightarrow S_7 - S_7 \rightarrow S_2 \rightarrow S_1 \rightarrow S_0$ is commonly used for each sampling cycle pair, namely switching cycle T_{sw}, when the reference voltage vector $\vec{V}_{ref}$ is located inside the sector I. This strategy creates the

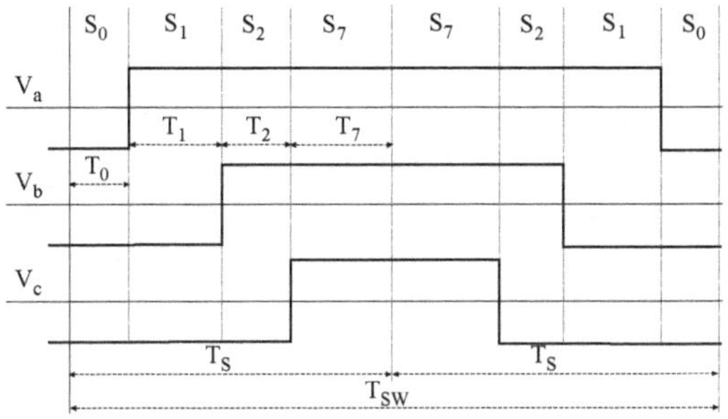

Fig. 4.4: Switching patterns generated by C-SVPWM considering $\vec{V}_{ref}$ in sector I.

switching pattern of Fig. 4.4. In this technique all three inverter legs are switched for all the time during the fundamental cycle T_1[2)] and there exists no switching absence interval. Hence, the method is called Continuous SVPWM (C-SVPWM), or "Three Phase Modulation".

The freedom of apportioning the zero voltage vector duty cycle $T_{0/7} = T_0 + T_7$ between the two zero switching S_0 and S_7 can be used to improve the performance of modulation. For this purpose an apportioning factor γ is defined in [66]:

$$\gamma = \frac{T_7}{T_0 + T_7} = \frac{T_7}{T_{0/7}}. \tag{4.6}$$

In the conventional C-SVPWM introduced firstly in [61], γ is simply set to 0.5, so the zero voltage vector duty cycle is equally divided between $\vec{V}_{0(nnn)}$ and $\vec{V}_{7(ppp)}$. Research works such as [66] have confirmed that the conventional SVPWM with $\gamma = 0.5$ is a fairly good choice for practical applications and can achieve almost the best results from the AC current waveform quality point of view.

In the second sector of the voltage hexagon, the zero switching states can be placed either in the same order as in sector I, i. e. S_0 at the sides of the switching cycle and S_7 in the middle, which is named C-SVPWM1, or in reverse order, S_0 in the middle and S_7 at the sides which is called C-SVPWM2. The switching patterns of both types are shown in table 4.1.

2) It should not be confused with the duty cycle of $\vec{V}_1$.

Voltage Sector					
I	II	III	IV	V	VI
Switching pattern of C-SVPWM1					
01277210	03277230	03477430	05477450	05677650	01677610
Switching pattern of C-SVPWM2					
01277210	72300327	03477430	74500547	05677650	76100167

Table 4.1: Switching patterns of C-SVPWM1 and C-SVPWM2.

Both types have the same locally (with respect to sampling cycle) averaged AC quantities, but the spectra of AC waveforms resulting from these types differ from each other to some extent. Fig. 4.5 shows the spectra of normalized phase current i_a and of the normalized common mode voltage v_{N0} derived from the Fourier analysis of waveforms simulated by the SABER simulator program

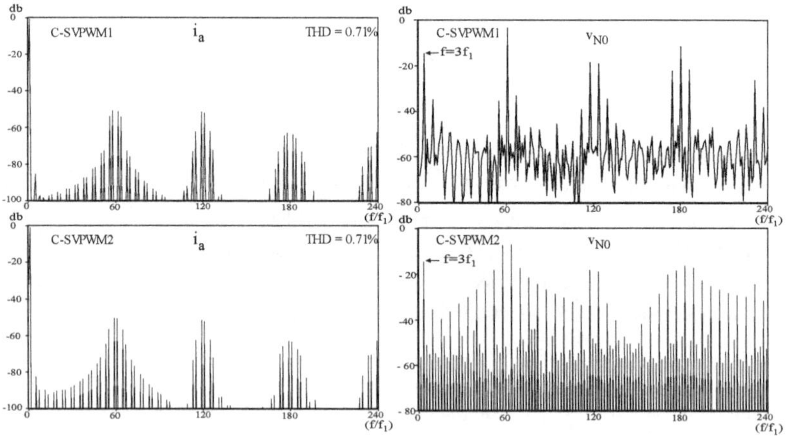

Fig. 4.5: Normalized phase current and common mode voltage spectra for the types C-SVPWM1 and C-SVPWM2 ($m = 0.9$, $f_s = 120\,f_1$).

[38]. In the figure all axes are normalized to the respective base values and the given THD values are calculated by the Fourier analysing function of SABER. See appendix B for the base values and the expression used by SABER for the calculation of THD.

As can be seen on Fig. 4.5 the spectra have their first maximum around the switching frequency $f_{sw} = f_s/2$, $(f_s = 1/T_S)$. The spectra of waveforms generated by the second type C-SVPWM2 contain only odd number harmonics. It is also worth noting that the spectrum of v_{N0} of both types has a 3rd harmonic with high amplitude as the case of HI-SPWM technique. This is marked in Fig. 4.5(right-top and bottom) with $f = 3f_1$. The high amplitude of 3rd harmonic in the spectrum of v_{N0} suggests the similarity of C-SVPWM and HI-SPWM methods which is the subject of the following section.

4.3.1.1 Carrier based C-SVPWM

Fig. 4.6 shows the normalized waveforms of locally averaged phase voltage, $\bar{v}_a$, and common mode voltages, $\bar{v}_{N0}$, of SVPWM technique along with the normalized sinusoidal reference voltage, $v_{ref(a)}$, 3rd harmonic voltage, v_{3rd}, and harmonic injected reference voltage waveform of the HI-SPWM technique, $v_{ref(a)} + v_{3rd}$. The peak value of triangle signal in HI-SPWM is also normalized to the same base value as three phase reference voltages in SVPWM, i.e. $V_{dc}/2$. In this way the modulation index m becomes equivalent to the amplitude of the normalized sinusoidal reference voltage, $v_{ref(a)}$, for both techniques.

Since in the direct digital techniques the reference voltages are practically sampled at the beginning of the sampling cycle, but will be realized during this cycle, a delay time as much as $T_S/2$ is expected between the reference voltage waveform and the waveform of the output voltage. In the performed simulations, the voltage waveforms are averaged at the end of the sampling cycle. Therefore a delay time equal to T_S is seen between the reference and locally averaged waveforms. Considering this delay while carefully examining the HI-SPWM reference voltage waveform, $v_{ref(a)} + v_{3rd}$, and the SVPWM locally averaged output voltage waveform, $\bar{v}_a$, in Fig. 4.6, it can be clearly understood that by adding an unique zero sequence voltage to the three phase sinusoidal reference voltages, $V_{ref(a)}$, $V_{ref(b)}$ and $V_{ref(c)}$, the Harmonic Injected technique (HI-SPWM) can generate the same inverter output voltages as the space vector technique (SVPWM)

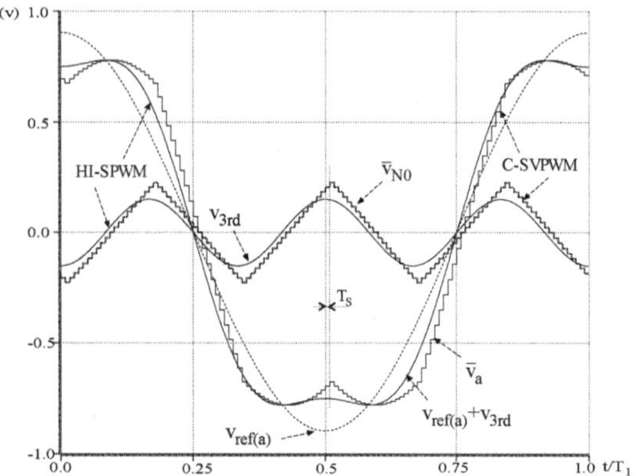

Fig. 4.6: Waveforms of continuous space vector (C-SVPWM) and harmonic injected (HI-SPWM) PWM techniques.

generates. In this way, generation of the SVPWM switching pattern shown in Fig. 4.4 with a simple analog comparator circuit becomes possible. This method is introduced in [67] as a hybrid PWM. It will be described here briefly.

Take the voltage sector I for example (Fig. 4.3). Since the line voltages V_{ab} and V_{bc} are equal to the DC bus voltage, V_{dc}, during duty cycles T_1 and T_2 (Fig. 4.4), T_1 and T_2 can be simply calculated from the volt-second balance principle over the sampling cycle T_S as follows:

$$\begin{cases} V_{dc}\,T_1 &= V_{ab}\,T_S \;\hat{=}\; V_{ref(ab)}\,T_S = (V_{ref(a)} - V_{ref(b)})\,T_S, \\ V_{dc}\,T_2 &= V_{bc}\,T_S \;\hat{=}\; V_{ref(bc)}\,T_S = (V_{ref(b)} - V_{ref(c)})\,T_S. \end{cases} \tag{4.7}$$

Normalizing these equations yields (see also appendix B):

$$\begin{cases} t_1 = \dfrac{v_{ref(ab)}}{2}, \\ t_2 = \dfrac{v_{ref(bc)}}{2}. \end{cases} \tag{4.8}$$

The normalized duty cycle of phases a, b and c, t_a, t_b and t_c, as shown in Fig. 4.7, are defined by:

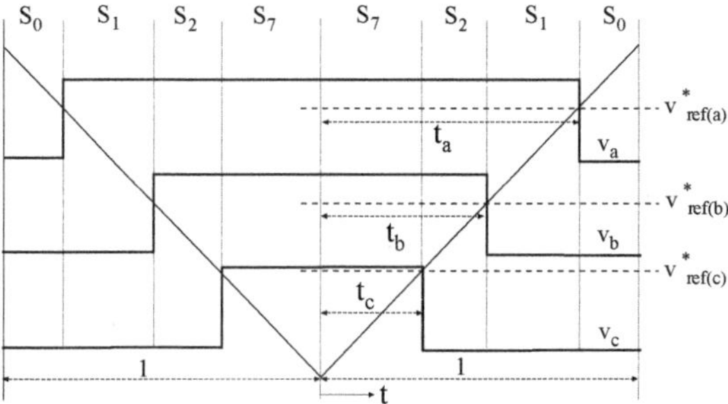

Fig. 4.7: Relationship of switching timing between SVPWM and SPWM(all durations are normalized to T_S).

$$\begin{cases} t_a &= \gamma\, t_{0/7} + t_2 + t_1, \\ t_b &= \gamma\, t_{0/7} + t_2, \\ t_c &= \gamma\, t_{0/7}, \end{cases} \tag{4.9}$$

where γ is the apportioning factor defined by (4.6). Introducing t_a, t_b and t_c into the normalized equation of triangle signal $v_t = 2t - 1$, $0 \le t \le 1$ (part with positive slope in Fig. 4.7) the new reference voltages are obtained as follows:

$$\begin{cases} v^*_{ref(a)} &= 2\,(\gamma\, t_{0/7} + t_2 + t_1) - 1, \\ v^*_{ref(b)} &= 2\,(\gamma\, t_{0/7} + t_2) - 1, \\ v^*_{ref(c)} &= 2\,(\gamma\, t_{0/7}) - 1. \end{cases} \tag{4.10}$$

Substituting $t_{0/7} = 1 - t_1 - t_2$ and, t_1 and t_2 from (4.8) in (4.10), the new normalized reference voltages are given by:

$$\begin{cases} v^*_{ref(a)} &= v_{ref(a)} + v_{zs}, \\ v^*_{ref(b)} &= v_{ref(b)} + v_{zs}, \\ v^*_{ref(c)} &= v_{ref(c)} + v_{zs}, \\ v_{zs} &= -[(1 - 2\,\gamma) + \gamma\, v_{ref(a)} + (1 - \gamma)\, v_{ref(c)}], \end{cases} \tag{4.11}$$

where $v_{ref(a)}$, $v_{ref(b)}$ and $v_{ref(c)}$ are the normalized primary sinusoidal three-phase reference voltages and $v^*_{ref(a)}$, $v^*_{ref(b)}$ and $v^*_{ref(c)}$ are the new ones obtained by adding the zero sequence voltage v_{zs} to the primary sinusoidal reference voltages.

Continuous Carrier-Based SVPWM						
VS	I	II	III	IV	V	VI
$2\,t_K$	$v_{ref(ab)}$	$v_{ref(ac)}$	$v_{ref(bc)}$	$v_{ref(ba)}$	$v_{ref(ca)}$	$v_{ref(ac)}$
$2\,t_{K+1}$	$v_{ref(bc)}$	$v_{ref(ba)}$	$v_{ref(ca)}$	$v_{ref(cb)}$	$v_{ref(ab)}$	$v_{ref(ac)}$
$v_{ref(min)}$	$v_{ref(c)}$	$v_{ref(c)}$	$v_{ref(a)}$	$v_{ref(a)}$	$v_{ref(b)}$	$v_{ref(b)}$
$v_{ref(mid)}$	$v_{ref(b)}$	$v_{ref(a)}$	$v_{ref(c)}$	$v_{ref(b)}$	$v_{ref(a)}$	$v_{ref(c)}$
$v_{ref(max)}$	$v_{ref(a)}$	$v_{ref(b)}$	$v_{ref(b)}$	$v_{ref(c)}$	$v_{ref(c)}$	$v_{ref(a)}$

Table 4.2: Dependence of t_K, t_{K+1}, $v_{ref(min)}$, $v_{ref(max)}$ and $v_{ref(mid)}$ on the number of voltage sector VS and on the line to neutral and line to line three-phase sinusoidal reference voltages for carrier-based continuous modulation.

The above expression describes the new three-phase reference voltages during the first 60° sector of the fundamental cycle (sector I of voltage hexagon in SVPWM). Performing the same for all other sectors, the following closed form can be concluded:

$$\begin{cases} v^*_{ref(a,b,c)} = v_{ref(a,b,c)} + v_{zs}, \\ v_{zs} = -[(1-2\,\gamma) + \gamma\, v_{ref(max)} + (1-\gamma)\, v_{ref(min)}] & \text{for} \quad 0 \le \gamma \le 1, \\ v_{zs} = -0.5\,(v_{ref(max)} + v_{ref(min)}) = 0.5\, v_{ref(mid)} & \text{for} \quad \gamma = 0.5, \end{cases} \tag{4.12}$$

where $v_{ref(max)}$, $v_{ref(min)}$ and $v_{ref(mid)}$ are maximum, minimum and middle values of normalized primary three-phase sinusoidal reference voltages, $v_{ref(a,b,c)}$, shown in Fig. 4.8(a). Fig. 4.8(b) illustrates the new reference voltage, $v^*_{ref(a)}$, and the zero sequence voltage, v_{zs}, as well as the locally averaged output, $\bar{v}_a$, and common mode voltages, $\bar{v}_{N0}$. Since with the new reference voltages the switching pattern of SVPWM (C-SVPWM1) can be generated by using the conventional Triangle-Comparison (TC) method, this technique is called carrier-based continuous space vector PWM. The dependence of t_K, t_{K+1}, $v_{ref(min)}$, $v_{ref(max)}$ and $v_{ref(mid)}$ on the number of voltage sector and on the sinusoidal reference voltages is shown in table 4.2 for all voltage sectors.

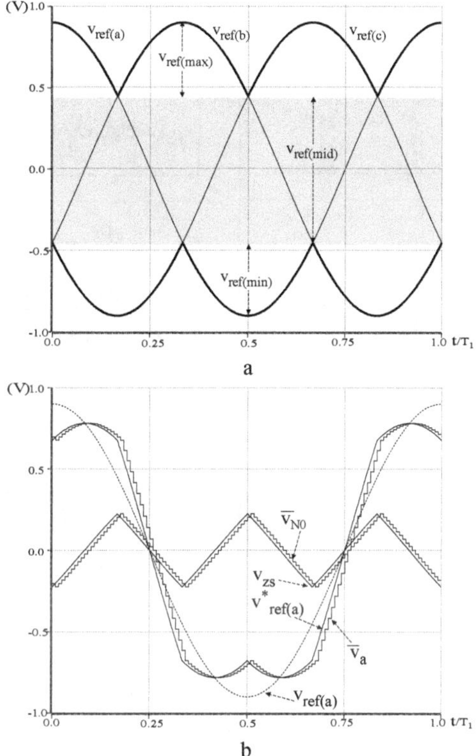

Fig. 4.8: Waveforms of carrier-based C-SVPWM.

Fig. 4.9 shows the normalized phase current and common mode voltage spectra for different well-known carrier based PWM. The resulting amplitude of 3rd harmonic component of v_{N0} of HI-SPWM technique corresponding to the amplitude of injected 3rd harmonic (Fig. 4.9(right-middle)) is close to that of C-SVPWM1 technique corresponding to the magnitude of added zero sequence voltage (Fig. 4.9(right-bottom)). The spectra of v_{N0} seem to be similar for both the HI-SVPWM and C-SVPWM1 techniques. Those of i_a are only different in the amplitude of harmonics around the switching frequency $f_s = 60\,f_1$ (Fig. 4.9(left-middle and bottom)). This is because of the different shape of added zero sequence signals in the HI-SVPWM and C-SVPWM1 techniques. The block diag-

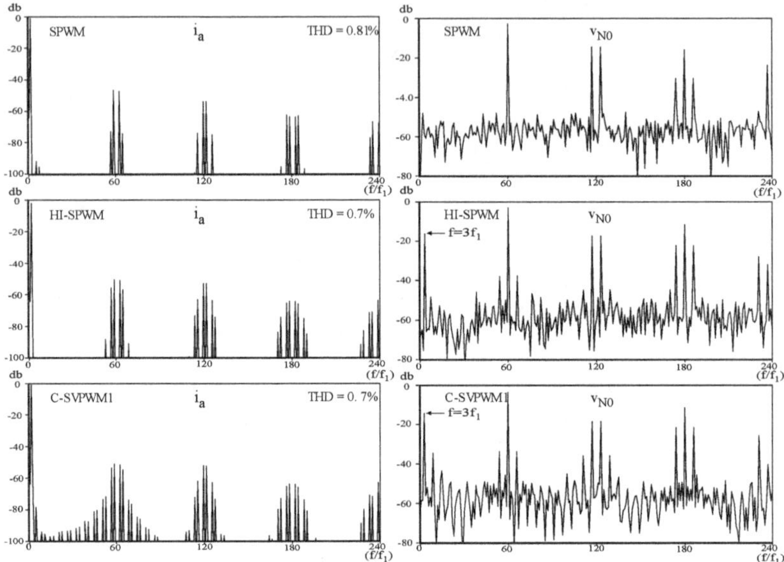

Fig. 4.9: Spectra of the normalized phase current and common mode voltage with different carrier based continuous PWM strategies ($m = 0.9$, $f_s = 120\, f_1$).

ram representation of the algorithm described by (4.12) is shown in Fig. 4.10. This algorithm is convenient for both digital and analog implementations. For an analog implementation, a three phase rectifier diode bridge with compensated voltage drops on diodes can be used for sorting.

4.3.2 Discontinuous Modulation

In contrary to the continuous modulation, the discontinuous modulation is a strategy which utilizes the aforementioned second and third degree of freedom of SVPWM in the discontinuous switching operation of inverter legs to optimize the inverter switching losses. In the discontinuous modulation techniques only one of the zero voltage vectors, $\vec{V}_7$ and $\vec{V}_0$, is switched in each switching cycle. This causes each phase to be alternately clamped to the positive or negative rail of DC link voltage during a so-called no switching interval. The preliminary advantage of this operation is the commutation of only two phase currents for each swit-

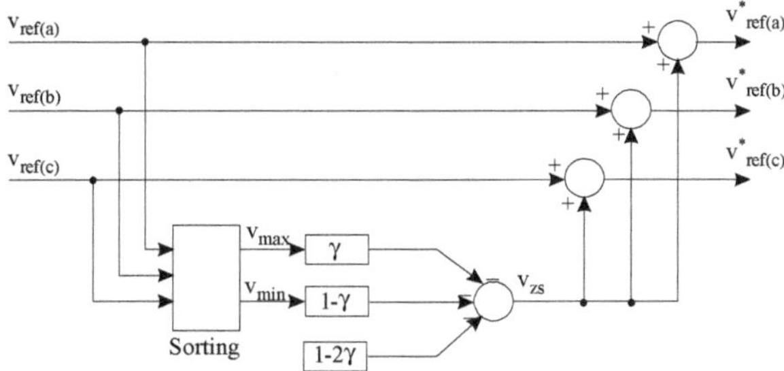

Fig. 4.10: Block diagram of carrier-based C-SVPWM.

ching cycle which reduces both main and auxiliary circuit loss to some extent. Different algorithms have been proposed in [70]-[75] to achieve this purpose. The algorithms using this strategy will be named in abbreviated form D-SVPWM in this work.

4.3.2.1 Discontinuous Modulation with One 120° Bus Clamped Interval

The simplest algorithm for generating a D-SVPWM switching pattern proposed in [70] is elimination of one of the zero vectors for the entire fundamental cycle. This strategy can suggest two cases: D7-SVPWM which uses only the zero voltage vector $\vec{V}_7$ and D0-SVPWM which uses only the voltage vector $\vec{V}_0$. Their switching patterns are given in table 4.3.

Switching Pattern for D7-SVPWM						
VS	I	II	III	IV	V	VI
SS	127721	327723	347743	547745	567765	167761
Switching Pattern of D0-SVPWM						
VS	I	II	III	IV	V	VI
SS	210012	230032	430034	450054	650056	610016

Table 4.3: Switching patterns of D7-SVPWM and D0-SVPWM (VS: Voltage Sector number; SS: Switching Sequence).

Although these algorithms are easy to implement, because of the higher temperature excursions on the power devices which create the zero switching state they have not been so widely employed.

Fig. 4.11 shows the spectra of the normalized phase current and common mode voltage waveforms for these algorithms. As can be seen on the spectra of v_{N0}, the Fourier description of the common mode voltage contains high amplitude at 3rd and DC components. The high amplitude of 3rd components suggests that these switching patterns can also be generated by the carrier based method, when a proper zero sequence voltage is added to the reference voltages. Consider the D7-SVPWM for example: elimination of zero voltage vector $\vec{V_0}$ from the switching sequence means $\gamma = 1$ in expression of v_{zs} given by (4.12). Substituting $\gamma = 1$ in this expression results in:

$$v_{zs_{(D7-SVPWM)}} = 1 - v_{ref(max)}, \tag{4.13}$$

and similarly for D0-SVPWM when substituting $\gamma = 0$ in this expression it follows:

$$v_{zs_{(D0-SVPWM)}} = -1 - v_{ref(min)}. \tag{4.14}$$

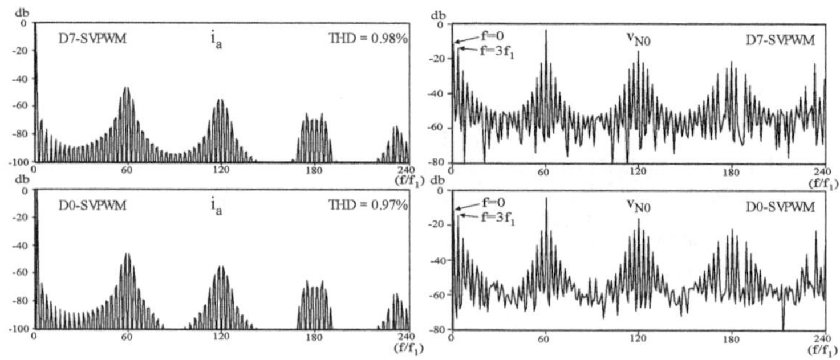

Fig. 4.11: Spectra of the normalized phase current and common mode voltage for D0-SVPWM and D7-SVPWM ($m = 0.9$, $f_s = 120\,f_1$).

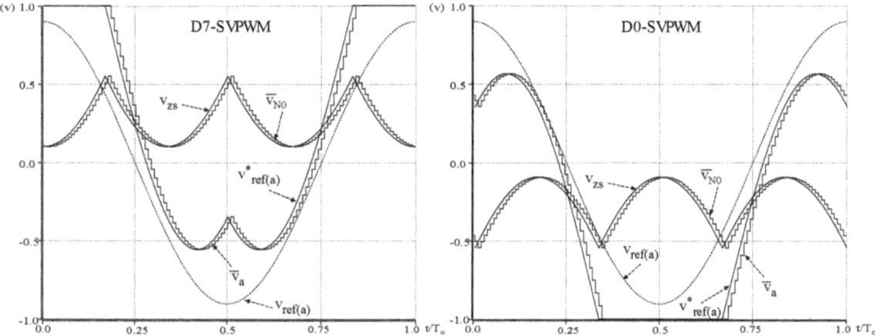

Fig. 4.12: Waveforms of the normalized zero sequence voltage v_{zs}, reference voltage $v^*_{ref(a)}$, locally averaged phase voltage $\bar{v}_a$ and common mode voltages $\bar{v}_{N0}$ for D7-SVPWM and D0-SVPWM ($m = 0.9,\ f_s = 120\,f_1$).

Fig. 4.12 illustrates the zero sequence voltage v_{zs} and the resulting reference voltage $v^*_{ref(a)}$ for both algorithms. In this figure the locally averaged phase voltage, $\bar{v}_a$, and common mode voltage, $\bar{v}_{N0}$, are shown too. The zero sequence voltage v_{zs} has a DC component which appears as a DC component in the spectrum of common mode voltage, v_{N0}, shown in Fig. 4.11(right). The reference voltage waveforms indicate that each phase has a 120° bus clamped interval in which the respective switches are left on/off. These techniques show advantage in reduction of number of switching actions which results less switching losses.

4.3.2.2 Discontinuous Modulation with Two 60° Bus Clamped Intervals

Since the switching losses of devices depend on the turn-on/off current, it is preferable to divide the 120° bus clamped interval of each phase into two 60° intervals and place them in the vicinity of the phase current peaks. Consequently, the freedom of choosing the zero voltage vector in SVPWM is utilized for minimizing the switching losses.

Suppose first the phase currents are in phase with the phase voltage, so that the current hexagon can be plotted inside the voltage hexagon as shown gray in Fig. 4.13(a). For the subsector Ia ($\vec{v}_{ref}$ in voltage sector I and $\vec{i}$ in current sector a), the switching pattern of D7-SVPWM is preferably used to avoid the

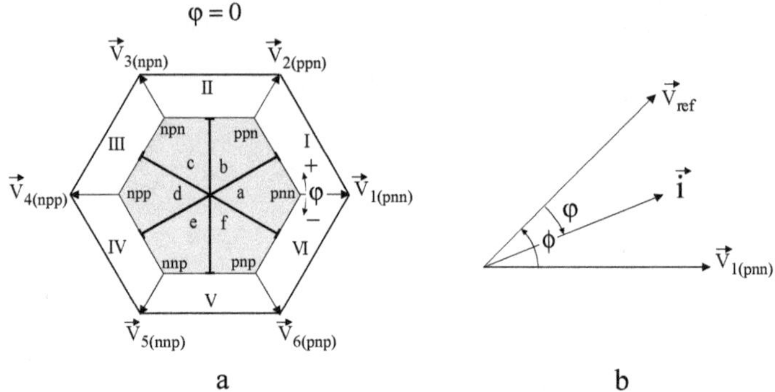

Fig. 4.13: a) Voltage and current hexagons when $\varphi = 0$ and b) definition of ϕ and φ.

Discontinuous Modulation with Two 60° Bus Clamped Intervals						
ML-SVPWM for $-30^o < \varphi < 30^o$						
VS	I		II		III	
φ	$- \cdots 0 \cdots +$		$- \cdots 0 \cdots +$		$- \cdots 0 \cdots +$	
CS	a	b		c		d
I_M	$i_{a(max)}$	$i_{c(min)}$		$i_{b(max)}$		$i_{a(min)}$
SS	127721	210012	230032	327723	347743	430034
v_{zs}	$1 - v_{ref(a)}$	$-1 - v_{ref(c)}$		$1 - v_{ref(b)}$		$-1 - v_{ref(a)}$
VS	IV		V		VI	
φ	$- \cdots 0 \cdots +$		$- \cdots 0 \cdots +$		$- \cdots 0 \cdots +$	
CS	d	e		f		a
I_M	$i_{a(min)}$	$i_{c(max)}$		$i_{b(min)}$		$i_{a(max)}$
SS	450054	547745	567765	650056	610016	167761
v_{zs}	$-1 - v_{ref(a)}$	$1 - v_{ref(c)}$		$-1 - v_{ref(b)}$		$1 - v_{ref(a)}$

Table 4.4: Dependence of switching pattern and v_{zs}, on the number of voltage and current sectors for ML-SVPWM when $-30^o < \varphi < 30^o$(VS: Voltage Sector number, CS: Current Sector number, I_M: Current Peak and SS: Switching Sequence).

switching actions in phase a, because the current of this phase has the absolutely largest magnitude within the current sector a. Analogous, for the subsector Ib the switching pattern of D0-SVPWM is chosen to avoid the switching actions in phase c. With the analogous analysis for all the other subsectors, the switching pattern of table 4.4 is obtained, which is called Minimum-Loss SVPWM (ML-SVPWM).

It should be also noted that in order to generate the switching pattern of each subsector using the carrier based method, the respective zero sequence voltage given for v_{zs} should be added to the sinusoidal reference voltages. Every segment of v_{zs} is composed of the zero sequence voltage of the D7/0-SVPWM technique used in that segment.

For the reactive or capacitive loads, the current hexagon shown gray in Fig. 4.13 turns anti-clockwise or clockwise respectively. Accordingly the single vertical lines under the φ measure in the table 4.4 corresponding to the current sectors move towards the right side or left side of the table. The double vertical lines corresponding to the voltage sectors remain unmoved. In this way the bus clamped intervals are always kept on the current peaks.

This algorithm is applicable to the range of φ confined to $[-\frac{\pi}{6}, \frac{\pi}{6}]$. The boundaries of this range are the angles at which the single vertical lines reach the double vertical lines in the table. For heavy reactive or capacitive loads, $-\frac{\pi}{3} < \varphi < -\frac{\pi}{6}$ and $\frac{\pi}{6} < \varphi < \frac{\pi}{3}$, the phase current with the absolutely largest magnitude has to be switched partly in vicinity of current peaks for transition between the non-zero vectors. So thereafter, the rows corresponding to v_{zs} and SS move no longer and switching actions at the phase current with the second largest magnitude are thereby avoided. Tables 4.5 and 4.6 show the switching patterns related to these ranges of φ.

This algorithm minimizes the switching losses for a wide range of load power factor variation. Figure 4.14 illustrates the zero sequence and reference voltages of optimized ML-SVPWM for different phase angles of inverter output current.

The angle ψ in these figures is described in the following of this section.

Discontinuous Modulation with two 60° Bus Clamped Intervals						
ML-SVPWM for heavy reactive loads $30^o < \varphi < 60^o$						
φ	$30^o \cdots \rightarrow \cdots 60^o$					
CS	a	b	c	d	e	f
I_M	$i_{a(max)}$	$i_{c(min)}$	$i_{b(max)}$	$i_{a(min)}$	$i_{c(max)}$	$i_{b(min)}$
VS	I	II	III	IV	V	VI
SS	127721	230032	347743	450054	567765	610016
v_{zs}	1-$v_{ref(a)}$	-1-$v_{ref(c)}$	1-$v_{ref(b)}$	-1-$v_{ref(a)}$	1-$v_{ref(c)}$	-1-$v_{ref(b)}$

Table 4.5: Dependence of switching pattern and v_{zs}, on the number of voltage and current sectors for ML-SVPWM when $30^o < \varphi < 60^o$ (VS: Voltage Sector number, CS: Current Sector number, I_M: Current Peak and SS: Switching Sequence).

Discontinuous Modulation with two 60° Bus Clamped Intervals						
ML-SVPWM for heavy capacitive loads $-60^o < \varphi < -30^o$						
φ	$-60^o \cdots \leftarrow \cdots - 30^o$					
CS	b	c	d	e	f	a
I_M	$i_{c(min)}$	$i_{b(max)}$	$i_{a(min)}$	$i_{c(max)}$	$i_{b(min)}$	$i_{a(max)}$
VS	I	II	III	IV	V	VI
SS	210012	327723	430034	547754	650056	167761
v_{zs}	-1-$v_{ref(c)}$	1-$v_{ref(b)}$	-1-$v_{ref(a)}$	1-$v_{ref(c)}$	-1-$v_{ref(b)}$	1-$v_{ref(a)}$

Table 4.6: Dependence of switching pattern and v_{zs}, on the number of voltage and current sectors for ML-SVPWM when $-60^o < \varphi < -30^o$ (VS: Voltage Sector number, CS: Current Sector number, I_M: Current Peak and SS: Switching Sequence).

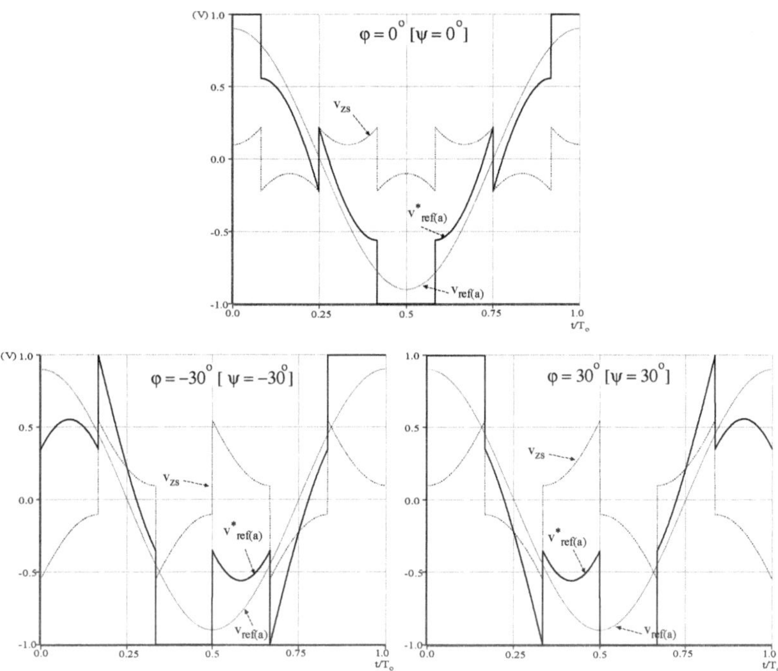

Fig. 4.14: Zero sequence and reference voltages for different phase angle of inverter current φ (m=0.9).

A careful examination of zero sequence voltages v_{zs} given in tables 4.4, 4.5 and 4.6 along with their waveforms in Fig. 4.14 indicates that the shape difference between these zero sequence waveforms arises from the 30° phase difference between their 60° segments. To create a generalized zero sequence voltage for the ML-SVPWM technique, a modulation angle ψ is defined as shown in Fig. 4.15(a).

For convenience of the algorithm, the origin of ψ is placed at $(-\frac{\pi}{6})$. Each 60° segment of the zero sequence voltage v_{zs} is obtained from the difference between the parts of $v_{ref(max)}$ or $v_{ref(min)}$ located in that segment and the saturation line (± 1). For $\psi = 0$, the voltages with the absolutely largest magnitude are the parts

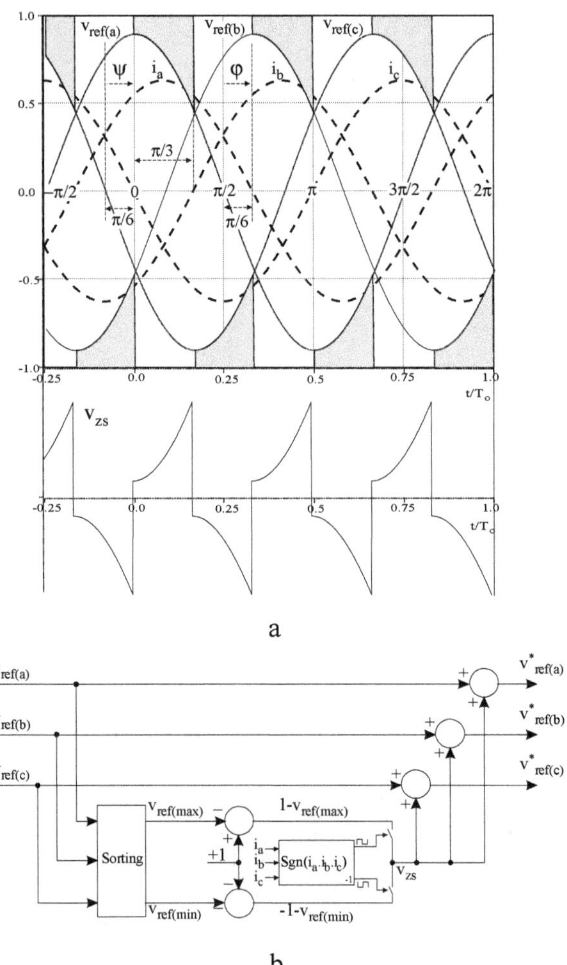

Fig. 4.15: a) ML-SVPWM zero sequence signal generation and b) implementation block diagram for the range $-\frac{\pi}{6} < \varphi < \frac{\pi}{6}$.

of $v_{ref(max)}$ or $v_{ref(min)}$ lying in the 60° segments. These segments are shifted lagging or leading by $\psi \neq 0$. In a similar modulation algorithm proposed only for the range $\varphi \in [-\frac{\pi}{6}, \frac{\pi}{6}]$ in [74] the origin of ψ is located at $(-\frac{\pi}{3})$ which is not an appropriate choice. By placing the origin of the modulation angle ψ at $-\frac{\pi}{6}$, ψ becomes equal to φ for this range. Hence, the segments of v_{zs} can be shifted according to the phase angle of output currents in this range. Also, further examination of the 60° segments of v_{zs} and the phase current waveforms i_a, i_b and i_c in Fig. 4.15(a) reveals that the zero sequence voltage segments are the intervals in which none of the phase currents changes its direction. Therefore the algorithm can be generalized for the mentioned range as follows: for the intervals, in which the sign of product of the phase currents is positive, the D7-SVPWM is selected and its zero sequence signal $(1 - v_{ref(max)})$ is added to the sinusoidal reference voltages. Analogous for the intervals in which this sign is negative, the zero sequence signal of D0-SVPWM, $(-1 - v_{ref(min)})$, is added to the sinusoidal reference voltages. As aforementioned, for the strong inductive and capacitive loads with the current phase angle beyond this range, the zero sequence segments shifting is stopped at $\varphi = 30^o$ or $\varphi = -30^o$. The zero sequence voltage for $\varphi > 30^o$ or $\varphi < -30^o$ remains the same as that of $\varphi = 30^o$ or $\varphi = -30^o$ respectively. This algorithm minimizes the switching losses for any load power factor in the range of $[-\frac{\pi}{3}, \frac{\pi}{3}]$ and therefore is quite attractive in applications with variable load power factor. The implementation of this algorithm as block diagram is shown in Fig. 4.15(b) for the range $[-\frac{\pi}{6}, \frac{\pi}{6}]$.

Figure 4.16 shows the spectra of the phase current and common mode voltage obtained from the modulation of reference voltages of Fig. 4.14. As can be seen from this figure, the shape of the reference voltage significantly affects the spectra of the phase current and common mode voltage.

Another discontinuous modulation technique with two 60° bus clamped intervals is given in table 4.7. This technique is used with the ARC-QW-ARC circuit and was described in chapter 2. In this technique the turn-on times of the on-going switches are synchronized. For this reason this technique which is usually used in non-pole soft commutated circuits is named Synchronous Turn-on SVPWM (ST-SVPWM) in this work.

Comparing the switching pattern of this table with that given in table 4.4, one can recognize that the zero sequence voltage as well as the locally averaged

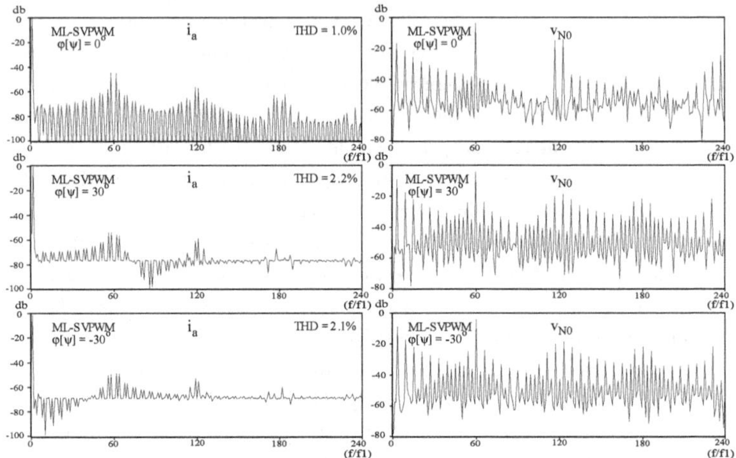

Fig. 4.16: Spectra of phase current and zero sequence voltage for different phase angle of inverter output current φ *(*$m = 0.9,\ f_s = 120\,f_1$*).*

Discontinuous Modulation with Two 60° Bus Clamped Intervals						
Synchronous Turns-on SVPWM (ST-SVPWM)						
VS	I		II		III	
CS	a	b		c		d
SS	721	012	032	723	743	034
VS	IV		V		VI	
CS	d	e		f		a
SS	054	745	765	056	016	761

Table 4.7: Dependence of switching pattern on the number of voltage and current sectors for ST-SVPWM when $-30^o < \varphi < 30^o$ *(*VS*: Voltage Sector number,* CS*: Current Sector number and* SS*: Switching Sequence).*

waveforms of the both techniques should be the same. The shifting algorithm of the zero sequence segments is also applicable to this technique. In contrary to the ML-SVPWM technique, the switching and sampling cycles (T_{sw} and T_S resp.) are equal in this technique. This results in higher harmonic contents for the AC waveforms. Fig. 4.17 shows the spectra of phase current and common

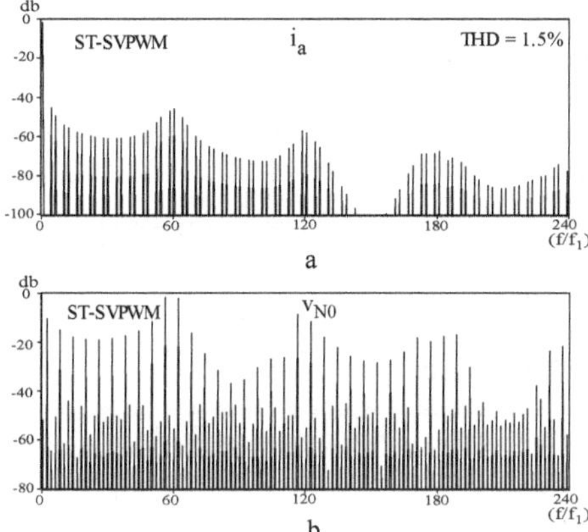

Fig. 4.17: ST-SVPWM ($\psi = 0^o$, $m = 0.9$; $f_s = 60\,f_1$): spectra of a) phase current and b) common mode voltage.

mode voltage for this technique in the same switching frequency, f_{sw}, as used for obtaining the spectra of other techniques. Despite the spectrum of phase current contains only odd number harmonics using this technique, total harmonic distortion (THD) of the phase current is higher because the reference voltages is sampled only once per switching cycle. (See also Fig.4.16 at $\psi = 0$)

4.3.2.3 Discontinuous Modulation with four 30° Bus Clamped Intervals

Swapping the places of the switching sequences of subsectors in each voltage sector of the table 4.4, a discontinuous modulation technique is found with four 30° bus clamped intervals distributed symmetrically over the fundamental cycle [76, 81]. Table 4.8 shows the switching pattern of this technique called D-SVPWM4.

This switching pattern can be generated with the carrier based method described in previous section by setting $\psi = \pm 60^o$. Figure 4.18 shows the waveforms of the reference and zero sequence voltages and the spectra of phase current and common mode voltage for this technique. The advantage of this technique is the symmetrical distribution of no switching area over the whole fundamental

Discontinuous Modulation D-SVPWM4 with four 30° Bus Clamped Intervals				
VS	I		II	
SS	210012	127721	327723	230032
v_{zs}	$-1-v_{ref(c)}$	$1-v_{ref(a)}$	$1-v_{ref(b)}$	$-1-v_{ref(c)}$
VS	III		IV	
SS	430034	347743	547745	450054
v_{zs}	$-1-v_{ref(a)}$	$1-v_{ref(b)}$	$1-v_{ref(c)}$	$-1-v_{ref(a)}$
VS	V		VI	
SS	650056	567765	167761	610016
v_{zs}	$-1-v_{ref(b)}$	$1-v_{ref(c)}$	$1-v_{ref(a)}$	$-1-v_{ref(b)}$

Table 4.8: Dependence of switching pattern and v_{zs} on the number of voltage sector for D-SVPWM4 (VS: Voltage Sector number and SS: Switching Sequence).

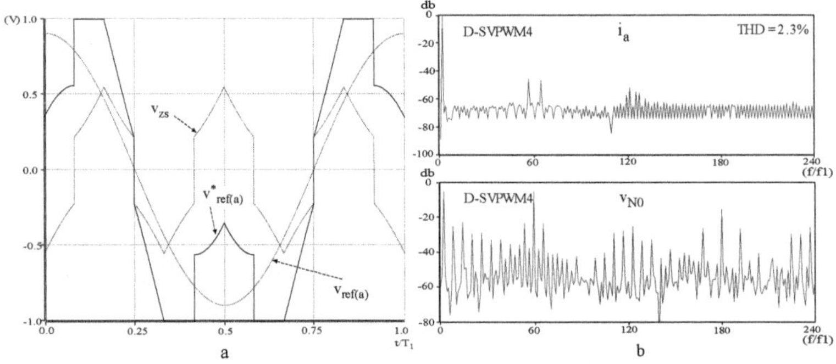

Fig. 4.18: D-SVPWM4 ($\psi = 60^o$, $m = 0.9$; $f_s = 120\, f_1$): a) the waveforms of modulation and zero sequence voltages and b) spectra of the phase current and common mode voltage.

cycle causing the switching losses of the inverter to become independent from the current phase angle which is of course due to switching actions in vicinity of current peaks sub-optimal.

Another idea is to arrange the switching sequence of subsectors of each voltage sector in such a way that the zero voltage vector changes every 30° instead of every 60° as in case of the aforementioned switching patterns. In order to have the zero vectors changing in each 30° interval, the zero sequence voltage should be also changed in each 30° interval. This arrangement results in a new switching pattern shown in table 4.9 and called D-SVPWM5.

Discontinuous Modulation D-SVPWM5 with four 30° Bus Clamped Intervals				
VS	I		II	
SS	127721	210012	327723	230032
v_{zs}	$1 - v_{ref(a)}$	$-1 - v_{ref(c)}$	$1 - v_{ref(b)}$	$-1 - v_{ref(c)}$
VS	III		IV	
SS	347743	430034	547745	450054
v_{zs}	$1 - v_{ref(b)}$	$-1 - v_{ref(a)}$	$1 - v_{ref(c)}$	$-1 - v_{ref(a)}$
VS	V		VI	
SS	567765	650056	167761	610016
v_{zs}	$1 - v_{ref(c)}$	$-1 - v_{ref(b)}$	$1 - v_{ref(a)}$	$-1 - v_{ref(b)}$

Table 4.9: Dependence of switching pattern and v_{zs} on the number of voltage sector for D-SVPWM5 ($\varphi = 0^o$) with four 30° bus clamped interval, 30° zero sequence voltage and zero vector changing segments (VS: Voltage Sector number and SS: Switching Sequence).

Fig. 4.19 shows the waveform of reference voltage, $v^*_{ref(a)}$, and of the zero sequence voltage, v_{zs}, as well as the spectra of the phase current, i_a, and of the common mode voltage, v_{N0}, for this switching pattern. Since the 30° bus clamped intervals are placed near to the phase current peaks, the switching losses are also sub-optimized. This new switching pattern is worthy, because changing the zero vector in every 30° interval results in lower temperature excursions on the power devices which create the zero vectors.

4.3.3 Influence of Different Carrier based SVPWM Techniques on the DC Bus Utilization and Linearity Range of Modulation

Apart from the influence of node voltage transition durations on the DC bus utilization and on the linearity range of the modulation in soft switching conver-

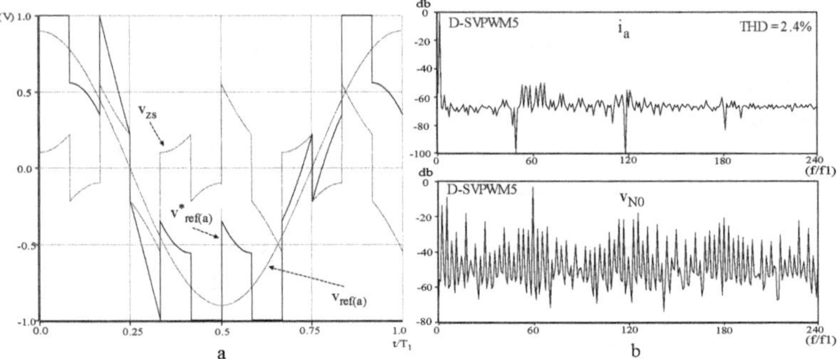

Fig. 4.19: D-SVPWM5 ($\varphi = 0^o$, $m = 0.9$; $f_s = 120\, f_1$): a) the waveforms of reference and zero sequence voltages and b) spectra of the phase current and common mode voltage.

ters, a subject to be discussed in chapter 5, different SVPWM techniques have no considerable effect on the linearity range of the modulation. Several simulations are carried out with these switching patterns to show this fact. The modulated voltages are than synthesized with the fast Fourier transform of SABER simulator. The amplitude of fundamental component of modulated phase voltage, normalized to $V_{dc}/2$, is explored and shown in Fig. 4.20 as a measure for the

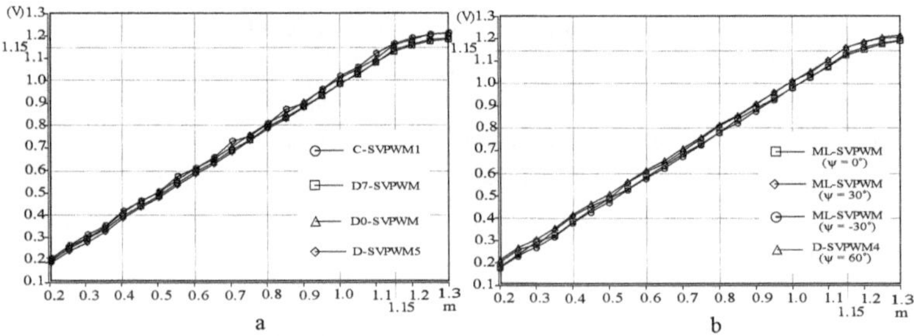

Fig. 4.20: Normalized amplitude of fundamental component of phase voltage versus modulation index for different carrier based SVPWM techniques.

DC bus utilization of different SVPWM techniques. The discontinuous switching patterns ML-SVPWM and D-SVPWM4 slightly affect the DC bus utilization of converters, but the linearity range of modulation as prerequisite for decoupled controller design was not outlined.

4.4 Influence of the SVPWM Techniques on the Inverter Energy Losses

To show the superiority of the SVPWM employing the discontinuous switching patterns over the classic continuous SVPWM techniques loss calculations and computer simulation have been performed. The losses will be calculated both locally (with respect to one sampling cycle for conduction loss and with respect to one switching cycle for switching losses) and globally (with respect to one fundamental cycle). The local conduction energy loss of each device is normalized to the conduction energy loss of that device when conducting the current maximum for one sampling cycle. Thereby the dependence of conduction energy loss on the sampling cycle, type of device and modulation index can be eliminated. This independence is realized for the local switching energy losses by normalizing these losses to the device switching energy losses at current maximum. The global energy losses of each technique are further normalized to the energy losses of C-SVPWM technique at maximum linear modulation index to show the ratio of global energy losses for respective technique.

4.4.1 Conduction Energy Losses

Conduction energy loss of the devices of each inverter bridge leg depend on the relative duty cycle of that leg called phase duty ratio. This duty ratio corresponds to the duty ratio (normalized duty cycle) of upper switch when the phase current is positive and to the duty ratio of upper diode when the phase current is negative (see the directions shown in Fig. 4.1). The dependence of the phase duty ratios t_a, t_b and t_c to the voltage vector duty ratios is given in equation (4.9) for sector I of the voltage hexagon. In the same manner as described in section 4.3.1.1 the duty ratio of each phase can be obtained for other voltage sectors, too. Introducing the respective voltage vector duty ratios, t_K and t_{K+1}, given in table 4.2 to t_a, t_b and t_c, the phase duty ratios are expressed by the reference voltages for all voltage (sub)sectors. Table 4.10 gives the duty ratio of phase a, t_a,

<table>
<tr><th colspan="6">Duty Ratio of Phase a, t_a</th></tr>
<tr><th>VS</th><th>C-SVPWM1</th><th>ML-(ψ=30)</th><th>ML-(ψ=-30°)</th><th>ML-(ψ=0°)</th><th>D-SVPWM5</th></tr>
<tr><td rowspan="2">I</td><td rowspan="2">$t_1+t_2+t_7=$
$\gamma+$
$(1-\gamma)\,v_{ref(ac)}/2$</td><td rowspan="2">1
1</td><td rowspan="2">$t_1+t_2=$
$v_{ref(ac)}/2$</td><td>1
1</td><td>1
1</td></tr>
<tr><td>$t_1+t_2=$
$v_{ref(ac)}/2$</td><td>$t_1+t_2=$
$v_{ref(ac)}/2$</td></tr>
<tr><td rowspan="2">II</td><td rowspan="2">$t_2+t_7=$
$[v_{ref(ac)}+$
$\gamma(2-v_{ref(bc)})]/2$</td><td rowspan="2">t_2
$v_{ref(ab)}/2$</td><td rowspan="2">$t_2+t_7=$
$1-$
$v_{ref(bc)}/2$</td><td>$t_2=$
$v_{ref(ab)}/2$</td><td>$t_2+t_7=$
$1-v_{ref(bc)}/2$</td></tr>
<tr><td>$t_2+t_7=$
$1-v_{ref(bc)}/2$</td><td>$t_2=$
$v_{ref(ab)}/2$</td></tr>
<tr><td rowspan="2">III</td><td rowspan="2">$t_7=$
$\gamma(1-v_{ref(ba)}/2)$</td><td rowspan="2">$t_7=$
$1-$
$v_{ref(ac)}/2$</td><td rowspan="2">0
0</td><td>$t_7=$
$1-v_{ref(ac)}/2$</td><td>$t_7=$
$1-v_{ref(ac)}/2$</td></tr>
<tr><td>0
0</td><td>0
0</td></tr>
<tr><td rowspan="2">IV</td><td rowspan="2">$t_7=$
$\gamma(1-v_{ref(ca)}/2)$</td><td rowspan="2">0
0</td><td rowspan="2">$t_7=$
$1-$
$v_{ref(ac)}/2$</td><td>0
0</td><td>$t_7=$
$1-v_{ref(ac)}/2$</td></tr>
<tr><td>$t_7=$
$1-v_{ref(ac)}/2$</td><td>0
0</td></tr>
<tr><td rowspan="2">V</td><td rowspan="2">$t_6+t_7=$
$[v_{ref(ab)}+$
$\gamma(2-v_{ref(cb)})]/2$</td><td rowspan="2">$t_6+t_7=$
$1-$
$v_{ref(ab)}/2$</td><td rowspan="2">$t_6=$
$v_{ref(bc)}/2$</td><td>$t_6+t_7=$
$1-v_{ref(ab)}/2$</td><td>$t_6+t_7=$
$1-v_{ref(ab)}/2$</td></tr>
<tr><td>$t_6=$
$v_{ref(bc)}/2$</td><td>$t_6=$
$v_{ref(bc)}/2$</td></tr>
<tr><td rowspan="2">VI</td><td rowspan="2">$t_6+t_1+t_7=$
$\gamma+$
$(1-\gamma)\,v_{ref(ab)}/2$</td><td rowspan="2">$t_6+t_1=$
$v_{ref(ac)}/2$</td><td rowspan="2">1
1</td><td>$t_6+t_1=$
$v_{ref(ac)}/2$</td><td>1
1</td></tr>
<tr><td>1
1</td><td>$t_6+t_1=$
$v_{ref(ac)}/2$</td></tr>
</table>

Table 4.10: Dependence of duty ratio of phase a, t_a, on the number of voltage sector VS for C-SVPWM1, ML-SVPWM ($\psi = 0$), ML-SVPWM ($\psi = 30^o$), ML-SVPWM ($\psi = -30^o$) and D-SVPWM5 techniques.

for the C-SVPWM1, ML-SVPWM and D-SVPWM5 techniques. The techniques D7-SVPWM, D0-SVPWM and D-SVPWM4 are excluded, because the energy losses of these techniques are well discussed in [77].

Table 4.10 represents also the duty ratio of switch S_{ap} (see Fig. 4.1) during the positive half cycle of phase current $I_a(\phi) = \hat{I}\cos(\phi - \varphi)$ (see also Fig. 4.13), when $-\frac{\pi}{2} + \varphi < \phi < \frac{\pi}{2} + \varphi$, and the duty cycle of diode D_{ap} during the negative half cycle of phase current I_a, when $\frac{\pi}{2} + \varphi < \phi < \frac{3\pi}{2} + \varphi$.

If the forward characteristic of the devices $V_{F,dev}(t)$ is approximated by:

$$V_{F,dev}(t) = V_{Fo,dev} + r_{F,dev}\, I_{F,dev}(t), \tag{4.15}$$

and the phase current is assumed constant during the sampling cycle T_S, then the local mean value of the conduction energy loss of switch S_{ap}, $E_{con(local),S_{ap}}(\phi)$, and of diode D_{ap}, $E_{con(local),D_{ap}}(\phi)$, are given by following equations:

$$\begin{cases} E_{con(local),S_{ap}}(\phi) = t_a(\phi)\left(V_{Fo,S}\, I_a(\phi) + r_{F,S}\, I_a^2(\phi)\right) & -\frac{\pi}{2} + \varphi < \phi \leq \frac{\pi}{2} + \varphi, \\ E_{con(local),S_{ap}}(\phi) = 0 & \frac{\pi}{2} + \varphi < \phi \leq \frac{3\pi}{2} + \varphi, \end{cases} \tag{4.16}$$

$$\begin{cases} E_{con(local),D_{ap}}(\phi) = 0 & -\frac{\pi}{2} + \varphi < \phi \leq \frac{\pi}{2} + \varphi, \\ E_{con(local),D_{ap}}(\phi) = t_a(\phi)\left(V_{Fo,D}\, |I_a(\phi)| + r_{F,D}\, I_a^2(\phi)\right) & \frac{\pi}{2} + \varphi < \phi \leq \frac{3\pi}{2} + \varphi, \end{cases} \tag{4.17}$$

where t_a is given for each voltage (sub)sector in table 4.10 as a function of normalized line reference voltages. The integration of these equations over the positive and negative half cycle of the phase current I_a results in the global mean value of the conduction energy loss of these devices, $E_{con(global),S_{ap}}(\phi)$ and $E_{con(global),D_{ap}}(\phi)$, as follows:

$$\begin{cases} E_{con(global),S_{ap}}(\phi) = \frac{1}{2\pi}\int_{-\frac{\pi}{2}+\varphi}^{\frac{\pi}{2}+\varphi} t_a(\phi)\left(V_{Fo,S}\, I_a(\phi) + r_{F,S}\, I_a^2(\phi)\right) d\phi, \\ E_{con(global),D_{ap}}(\phi) = \frac{1}{2\pi}\int_{\frac{\pi}{2}+\varphi}^{\frac{3\pi}{2}+\varphi} t_a(\phi)\left(V_{Fo,D}\, |I_a(\phi)| + r_{F,D}\, I_a^2(\phi)\right) d\phi. \end{cases} \tag{4.18}$$

As t_a is different for different SVPWM techniques, a relative change in conduction energy loss of devices can be expected for different techniques. The local and global mean value of the conduction energy loss of S_{ap} and of D_{ap} were calculated using the above equations by SABER. The local mean value of conduction energy loss of each device was normalized to the energy dissipated in the respective device when conducting the phase current maximum for one sampling cycle.

Figure 4.21(middle and bottom of a and b) shows the normalized local mean value of conduction energy losses of S_{ap} and D_{ap} for C-SVPWM1 and ML-SVPWM techniques. As can be seen from this figure the local mean value of the conduction energy loss of S_{ap} becomes maximum at $\phi = \pm 30^o$ when $\varphi = \pm 30^o$ (Fig. 4.21(a-middle)), because the duty ratio of S_{ap} in the C-SVPWM1 technique becomes maximum at the middle of voltage sectors I and VI. From Fig. 4.21(b-middle) shifting of the no switching intervals within $-60^o < \phi < 60^o$ according to tables 4.4, 4.5 and 4.6 can be also seen. After calculating the global mean value of conduction energy loss, this value is further normalized to the maximum of sum of global conduction energy losses of S_{ap} and D_{ap} of the C-SVPWM1 technique, which occurs at $\varphi = \pm 60^o$. The resulting waveforms shown in Fig. 4.21(a and b top) represent also the respective conduction energy loss ratio. With $\varphi = 0$, S_{ap} conducts the phase current for the major time during the positive half cycle because of overlapping the positive half cycle of phase voltage and current. With increase of the current phase angle (leading or lagging) and decrease of the overlapping parts of the positive half cycle of phase voltage and current the conduction duration of S_{ap} decreases, while that of D_{ap} increases. But note, that the turn-on duration of S_{ap} does not change with varying φ. It depends only on ϕ, m and the modulation technique used.

The sum of the global conduction energy losses of S_{ap} and D_{ap} for the continuous and discontinuous techniques with 60° or 30° bus clamped intervals are very close to each other. However, it will be very different for the D7-SVPWM and D0-SVPWM algorithms due to use of only one zero voltage vector. A big difference between the global conduction energy loss of the upper and bottom switches for the D7-SVPWM and D0-SVPWM algorithms is also expected. The same is true for the upper and lower diodes as confirmed in [77]. But, a very important point to note is that the overall conduction energy losses of the inverter main circuit (all main diodes and switches) do not vary much with the method used. This is true for any value of power factor and modulation index.

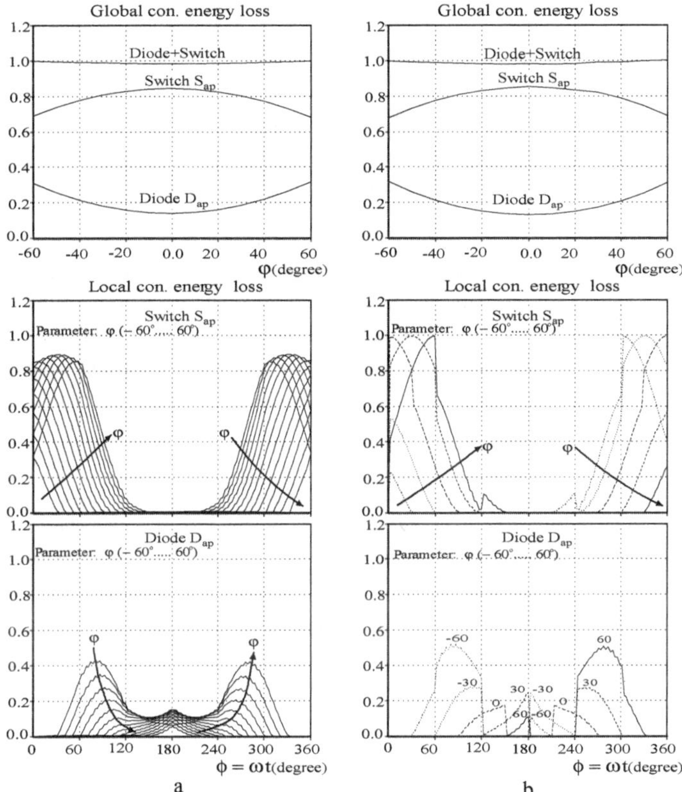

Fig. 4.21: Normalized local and global mean value of conduction energy loss of the top switch and diode at phase a *for a) C-SVPWM1(*$\gamma = 0.5$*) and b) ML-SVPWM with varying* φ *(both with* $m = 0.9$*).*

4.4.2 Switching Energy Losses

For the soft switching inverter, the switching energy losses of the main and auxiliary circuits are both important. Therefore a separately comparative evaluation for these losses is performed to show the advantages of the discontinuous methods for the soft switching inverters. The methods D7-SVPWM and D0-SVPWM are excluded because of their higher temperature excursion on the switches creating the zero vectors. The ST-SVPWM technique will also not be considered in the discussion of this section. The switching energy losses of this technique is twice as much as the ML-SVPWM technique at the same sampling frequency, because in this technique T_S equals T_{sw}.

4.4.2.1 Switching Energy Losses of Main Circuit

As described in chapter 2, turn-on actions of main switches in a zero voltage soft switching inverter are assumed to be lossless[3)] , because main switches always turn on while their antiparallel diode is conducting. Therefore, the only switching energy loss in the main circuit of a soft switching PWM inverter is assumed to be turn-off energy loss caused by switch-to-diode current commutations[4)] .

The turn-off energy loss caused by a sinusoidal current can be simply described by using the linear fall model of turn-off current given in [44]. Thus, the switching energy loss is assumed to be proportional to the square of the turn-off current (see also appendix A). Assumption of a linear dependence of the switching energy losses on the switched current (as made in [81]) cannot be true in case of soft switching inverters. As a comparative measure for turn-off energy loss a so-called Turn-off Energy Loss Indicator (TELI) is used. The TELI is calculated from the sum of squared instantaneous value of the phase current, I_a^2, at instants of switch turn-off actions at phase a.

Assuming the linear model of device fall current, the turn-off energy loss is given by (see appendix A):

$$E_{off} = \frac{I_{off}^2 \, t_f^2}{24\, C_S}, \tag{4.19}$$

where I_{off}, t_f and C_S are the turn-off current, turn-off time and the snubber capacitor paralleled to the switch, respectively. Normalization of this loss to the turn-off energy loss occurring at current maximum, eliminates the dependence of the expression given above on the type of devices and snubber capacitances, if a constant turn-off time is assumed. This assumption can be made, because

---

3) The gate losses of switches are small compared to conduction and switching losses. They are ignored in this discussion.

4) The losses caused by boost current in HW-topologies are also ignored in this discussion.

of comparative essence of our study[5]) . The normalized local turn-off energy loss indicator $\mathcal{E}_{off(local)}$ is therefore defined as follows:

$$\mathcal{E}_{off(local)} \hat{=} i_{off}^2. \tag{4.20}$$

Within the switching cycle T_{sw}, the current of phase a, I_a, is commutated once from the upper switch to the bottom diode and once from the bottom diode to the upper switch, when the current is positive[6]) . For the negative phase current the same is true for the bottom switch and upper diode. As mentioned earlier, the diode-to-switch current commutation causes no losses in the main circuit neglecting the boost current required in the HW-concept (see chapter 2). Therefore, when the phase current is assumed to be constant within the switching cycle, it is enough for the calculation of the normalized global turn-off energy loss indicator, $e_{off(global)}$, to add the square value of the normalized phase current, i_a, once per switching cycle. Thus, assuming a sufficiently high switching frequency, it results for the C-SVPWM technique:

$$\mathcal{E}_{\left(\substack{off(global)\\ c-svpwm}\right)} = 4\left[\sum_n \mathcal{E}_{off(local)}\right]_{\varphi}^{\frac{\pi}{2}+\varphi} \approx 4\int_{\varphi}^{\frac{\pi}{2}+\varphi} m^2\cos^2(\phi-\varphi). \tag{4.21}$$

In the above expression, the increase of the turn-off current caused by the auxiliary circuit for commutation of small phase currents in switch-to-diode current commutations is neglected for simplicity.

The expression of $\mathcal{E}_{off(global)}$ for the ML-SVPWM technique can be derived from (4.21) by subtracting the energy losses related to the bus clamped intervals from the expression given above. Considering the fact that the bus clamped intervals are always placed at the current peaks gives:

$$\begin{aligned}\mathcal{E}_{\left(\substack{off(global)\\ ml-svpwm}\right)} &= \mathcal{E}_{\left(\substack{off(global)\\ c-svpwm}\right)} - 4\left[\sum_n \mathcal{E}_{off(local)}\right]_{\varphi}^{\frac{\pi}{6}+\varphi} \\ &\approx \mathcal{E}_{\left(\substack{off(global)\\ c-svpwm}\right)} - 4\int_{\varphi}^{\frac{\pi}{6}+\varphi} m^2\cos^2(\phi-\varphi) \quad \text{for } -\frac{\pi}{6}\le\varphi\le\frac{\pi}{6}.\end{aligned} \tag{4.22}$$

[5]) Note, that exact calculation of the turn-off energy loss is not the target of this discussion. The Turn-off Energy Loss Indicator (TELI) is to consider as a measure for comparing the different SVPWM techniques.

[6]) With respect to the direction shown in Fig. 4.1.

Further normalization of the global turn-off energy loss to that of the C-SVPWM technique results in:

$$R_{ml-svpwm}(\varphi) = \frac{\mathcal{E}_{\binom{off(global)}{ml-svpwm}}}{\mathcal{E}_{\binom{off(global)}{c-svpwm}}} \approx 1 - \frac{\int_{\varphi}^{\frac{\pi}{6}+\varphi} \cos^2(\phi-\varphi)d\phi}{\int_{\varphi}^{\frac{\pi}{2}+\varphi} \cos^2(\phi-\varphi)d\phi} = \frac{2}{3} - \frac{\sqrt{3}}{2\,\pi} = 0.39 \quad \text{for } -\frac{\pi}{6} \leq \varphi \leq \frac{\pi}{6}, \tag{4.23}$$

where $R_{ml-vspwm}(\varphi)$ indicates the ratio of turn-off energy loss of the minimum-loss technique with $\psi = \varphi$ to turn-off energy loss of the C-SVPWM technique.

For the range ($\varphi < -\frac{\pi}{6}$ & $\varphi > \frac{\pi}{6}$) this ratio depends on φ. While considering $v^*_{ref(a)}(\phi) = -v^*_{ref(a)}(\pi + \phi)$, the ratio for this range can be calculated from:

$$R_{ml-svpwm}(\varphi) \approx 1 - \frac{2}{\pi} \int_{0\,[\frac{2\pi}{3}]}^{\frac{\pi}{3}\,[\pi]} \cos^2(\phi - \varphi)d\phi \quad \text{for} \quad \varphi > \frac{\pi}{6} \quad [\varphi < -\frac{\pi}{6}], \tag{4.24}$$

which leads to the results summarized in following expressions:

$$\begin{cases} R_{ml-svpwm}(\varphi) = \frac{2}{3} - \frac{\sqrt{3}}{2\,\pi} = 0.39 & \text{for} \quad -\frac{\pi}{6} \leq \varphi \leq \frac{\pi}{6}, \\ R_{ml-svpwm}(\varphi) = \frac{2}{3} - \frac{\sqrt{3}}{4\pi}\Big(\cos(2\varphi) + \sqrt{3}\sin(2\varphi)\Big) & \text{for} \quad \varphi > \frac{\pi}{6}, \\ R_{ml-svpwm}(\varphi) = \frac{2}{3} - \frac{\sqrt{3}}{4\pi}\Big(\cos(2\varphi) - \sqrt{3}\sin(2\varphi)\Big) & \text{for} \quad \varphi < -\frac{\pi}{6}. \end{cases} \tag{4.25}$$

Above expressions reveal the superiority of ML-SVPWM technique. By keeping $\psi = \varphi$, the highest reduction of turn-off energy loss (60%) can be achieved, if the variation of current phase angle φ is confined to the range $[-\frac{\pi}{6}, \frac{\pi}{6}]$ which is the typical PWM-VSI drives operation range. Note, that the turn-off energy loss can be reduced by as much as 60%, even though only 30% of the switching actions are eliminated. This is due to the elimination of the switching actions when the

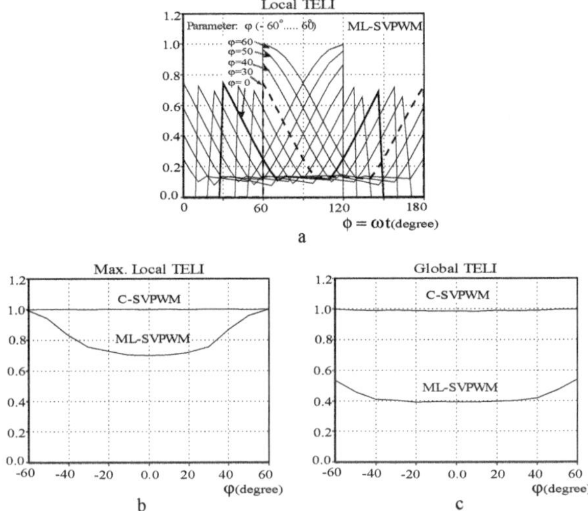

Fig. 4.22: Normalized local and global turn-off energy losses for C-SVPWM and ML-SVPWM techniques with different φ: a) local energy loss, b) maximum local energy loss and c) global energy loss.

phase current is at its maximum.

The proposed algorithm also provides the highest reduction in the turn-off energy loss for heavy reactive or capacitive loads, so that for $\varphi = \pm\frac{\pi}{3}$ a reduction of 48% can be expected. To confirm these results simulations are performed. The simulation results are shown in Fig. 4.22.

Fig. 4.22(a) illustrates the indicator of normalized local turn-off energy loss. For $\varphi < -\frac{\pi}{6}$ & $\varphi > \frac{\pi}{6}$ turn-off local energy loss increases in the voltage sector II (as shown in this figure) and also in the voltage sector V, because the absolute current maximum intervals start to leave the bus clamped intervals. The maximum of local turn-off energy loss roughly remains the same for the range $-\frac{\pi}{6} < \varphi < \frac{\pi}{6}$ as shown in Fig. 4.22(b). Also for this range a reduction of 60% can be seen in the normalized global turn-off energy loss in Fig. 4.22(c). Without continuous change of the reference voltages $v^*_{ref(a,b,c)}$ according to variation of φ, i.e. on-line adjustment of $\psi = \varphi$, this reduction cannot be always achieved. With a very good combination of the switching patterns given in tables 4.4, 4.5 and 4.6, the authors of [80] reached an average reduction of 48%.

The turn-off energy loss ratio functions of the techniques with four 30° bus clamped intervals, $R_{d-svpwm4}$ and $R_{d-svpwm5}$, can be derived in the same manner from the following general expression:

$$R_{d-svpwm}(\varphi) = 1 - \frac{1}{\pi}\sum_{n=1}^{4}\int_{\alpha_n}^{\alpha_n+\frac{\pi}{6}} \cos^2(\phi - \varphi)d\phi, \tag{4.26}$$

where α_n is the beginning of the nth bus clamped interval. Performing the same calculations results in:

$$R_{d-svpwm4}(\varphi) = \frac{2}{3}, \tag{4.27}$$

$$R_{d-svpwm5} = \frac{2}{3} - \frac{\sqrt{3}}{4\pi}\cos(2\varphi). \tag{4.28}$$

It is obvious that the turn-off energy loss of D-SVPWM4 technique are independent from the current phase angle, because the four bus clamped intervals are symmetrically distributed over the fundamental cycle. With the D-SVPWM5 technique, a maximum reduction of 47% for global turn-off energy loss can be expected when power factor is one. The strategy of shifting the bus clamped intervals can be also employed with the D-SVPWM5 technique to keep this ratio for the whole range of $\varphi \in [-\frac{\pi}{6}, \frac{\pi}{6}]$.

To show the effectiveness of D-SVPWM techniques over the whole range of modulation index, simulations are carried out at different modulation indices with $\psi = \varphi$ for ML-SVPWM and $\varphi = 0$ for others. In order to obtain the global energy loss indicator waveforms dependent on the modulation index, the local turn-off energy loss is normalised to that caused by the current maximum at the highest linear modulation index. The global turn-off energy loss, which is then calculated from the sum of these local values, is normalized again to the respective global turn-off energy loss of C-SVPWM at the highest linear modulation index. So the obtained global turn-off energy loss indicator will only depend on the modulation index and the SVPWM technique used. The results are shown in Fig. 4.23. In order to consider the turn-off current increase caused by the auxiliary

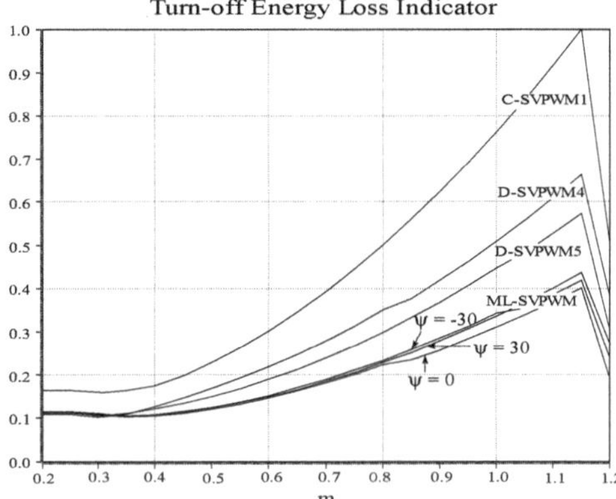

Fig. 4.23: Normalized global turn-off energy loss indicator versus modulation index using different SVPWM techniques.

circuit becoming active for the small current magnitudes in switch-to-diode current commutations in the simulations, in case the phase current is smaller than a quarter of the phase current amplitude at the highest linear modulation index, $\hat{I}_a(m = 1.15)/4$, the turn-off current is assumed to be equal to $\hat{I}_a(m = 1.15)/3$.

Figure 4.23 evidently indicates that the ML-SVPWM technique, only applicable to the commutated pole circuits (ACS-HW/QW-ARCP), allows an increase of the switching frequency twice as much as that is admissible for the classical C-SVPWM1 technique, for the switching losses are extremely reduced using this technique. This is while the circuit using the ML-SVPWM technique can still enjoy a few percent lower switching losses.

4.4.2.2 Energy Losses of the Auxiliary Circuit

The energy losses of the auxiliary circuit strongly depends on the topology of this circuit. But, since the purpose of this section is a comparative investigation of different SVPWM techniques, the effect of the SVPWM techniques on the auxiliary circuit energy losses can be investigated on a certain topology. For this purpose the ACS-HW-ARCP PWM inverter is selected. The following assumptions are made for simplicity of discussion:

- Both turn-on and turn-off actions of the auxiliary switches are lossless due to zero current switching conditions in the auxiliary circuit, so the auxiliary circuit energy losses contain only the conduction energy loss[7)].

- The increase and decrease of the auxiliary current are linear and have the same duration.

- The auxiliary current reaches a maximum as large as 1.5 times of the magnitude of the current being commutated, i. e. $I_{xa(max)}(\phi) = 1.5\,|I_a(\phi)|$, in the middle of transition duration.

- For switch-to-diode commutations with the phase current magnitude smaller than a quarter of the phase current maximum at $m = 1.15$, $|I_a(\phi)| < \hat{I}_a(m = 1.15)/4$, the auxiliary circuit is activated and increases the commutating current to $\hat{I}_a(m = 1.15)/3$. Thereby the auxiliary current maximum reaches the value $\left(1.5\,[\hat{I}_a(m = 1.15)/3 - |I_a(\phi)|]\right)$.

The conduction energy loss of the auxiliary switches, $E_{con(aux,S)}$, and of the auxiliary diodes, $E_{con(aux,D)}$, can be calculated in the same manner given by (4.16) and (4.17) as follows:

$$\begin{cases} E_{con(aux,S)}(\phi) = \left[V_{Fo,S}\,\dfrac{I_{xa(max)}(\phi)}{2} + r_{F,S}\left(\dfrac{I_{xa(max)}(\phi)}{2}\right)^2\right]\dfrac{4\,L_{aux}\,I_{xa(max)}(\phi)}{V_{dc}}, \\ E_{con(aux,D)}(\phi) = \left[V_{F,D}\,\dfrac{I_{xa(max)}(\phi)}{2} + r_{F,D}\left(\dfrac{I_{xa(max)}(\phi)}{2}\right)^2\right]\dfrac{4\,L_{aux}\,I_{xa(max)}(\phi)}{V_{dc}}. \end{cases} \tag{4.29}$$

Normalization of conduction energy loss of each device to the respective energy loss at the phase current maximum eliminates the dependence of these descriptions on the device type, L_{aux} and V_{dc}. The auxiliary circuit local energy losses is obtained by summing terms given above. The expressions of the auxiliary circuit global energy losses and from there the ratio of these losses to the energy losses related to C-SVPWM at the highest linear modulation index can be derived as explained for the turn-off energy losses in the last section. The auxiliary circuit normalized local and global energy losses were calculated by a SABER Program. The calculated results are shown in Fig. 4.24.

---

7) The gate losses of the auxiliary switches and ohmic loss of the auxiliary inductor are ignored.

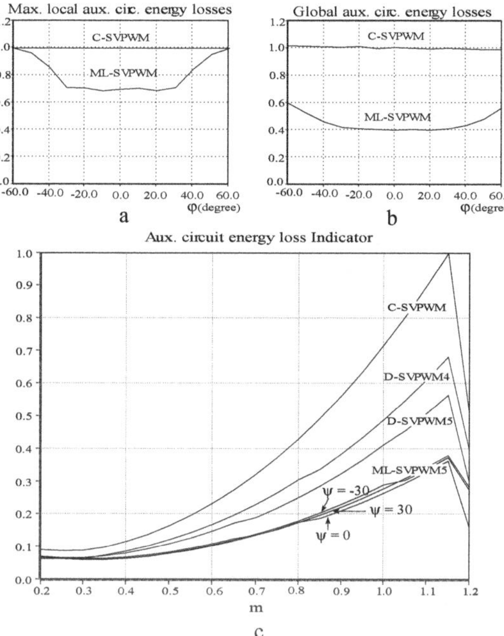

Fig. 4.24: Normalized energy losses of the auxiliary circuit: a) maximum of local energy losses, b) global energy losses ratio (with respect to energy losses of C-SVPWM at $\varphi = 0$) and c) energy loss ratio indicator (with respect to energy losses of C-SVPWM at $m = 1.15$) versus modulation index.

As it was expected, the results are similar to those obtained for the turn-off energy losses because the conduction energy losses of the auxiliary circuit also depend on the magnitude of the current being commutated. At low modulation indices, the auxiliary circuit will be activated during the most part of the fundamental period for both diode-to-switch and switch-to-diode current commutations due to the small amplitude of the phase current. It can be well observed from the shape of the energy loss indicator at modulation indices lower than 0.3 (Fig. 4.24(c)).

The results of this section and of the previous section as well, indicate that the ML-SVPWM and other D-SVPWM techniques are superior to C-SVPWM technique from the losses point of view. It is also worth recalling that among several three phase soft switching PWM inverters only the commutated pole concepts of AC side soft commutated PWM converters are able to operate with these techniques. In these circuits, the second and third degrees of freedom of the modulation are available for optimizing some criteria such as losses. In other circuits, these degrees of freedom have to be utilized to create a certain technique to meet the requirements of the auxiliary circuit.

4.5 Influence of the SVPWM Techniques on the Waveform Quality of AC Current

One of the significant points in comparison among the different modulation techniques is the waveform quality of output current which influences the size and costs of AC/EMI filter generating a large amount of volume and costs of the total assembly from one side and the losses in AC machine loads caused by higher order harmonics of AC output current from the other side. The waveform quality of the inverter output current is determined with current ripples caused by the deviation of the output current waveform from the current fundamental component waveform. This deviation appears because the inverter produces one of the voltage vectors $\vec{V}_K$ $(K = 0, \cdots, 7)$ at its output instead of the exact value of $\vec{V}_{ref}$. When the high frequency model of the load (AC machine) is approximated by an inductance L only, assuming the switching frequency high enough, current deviation vector, $\Delta\vec{I}_K(t)$, caused by switching of the voltage vector $\vec{V}_K$ at the inverter output can be described by:

$$\Delta\vec{I}_K(t) = \frac{1}{L}\left(\vec{V}_K - \vec{V}_{ref}(\phi)\right)(t - t_0). \tag{4.30}$$

This vector is expressed for simplicity in $\alpha\beta$-plane as follows:

$$\begin{cases} \Delta I_K^{\alpha}(t) & = \dfrac{1}{L}\left(\dfrac{2}{3}V_{dc}\cos\left((K-1)\dfrac{\pi}{3}\right) - \dfrac{m\,V_{dc}}{2}\cos(\phi)\right)(t-t_0), \\ & \qquad\qquad\qquad\qquad \text{for} \qquad K = 1,\cdots,6 \\ \Delta I_K^{\beta}(t) & = \dfrac{1}{L}\left(\dfrac{2}{3}V_{dc}\sin\left((K-1)\dfrac{\pi}{3}\right) - \dfrac{m\,V_{dc}}{2}\sin(\phi)\right)(t-t_0), \end{cases} \tag{4.31}$$

and

$$\begin{cases} \Delta I_K^{\alpha}(t) & = \dfrac{1}{L}\left(-\dfrac{m\,V_{dc}}{2}\cos(\phi)\right)(t-t_0), \\ & \qquad\qquad\qquad \text{for} \qquad K = 0,\,7 \\ \Delta I_K^{\beta}(t) & = \dfrac{1}{L}\left(-\dfrac{m\,V_{dc}}{2}\sin(\phi)\right)(t-t_0), \end{cases} \tag{4.32}$$

The waveform quality of the inverter output current is determined by the global (with respect to the fundamental cycle) per phase square RMS value (called later global RMS value) of current ripples, $\Delta I^2_{rms(global)}$, which can be calculated in $\alpha\beta$-plane from:

$$\begin{aligned} \Delta I^2_{rms(global)} &= \frac{1}{T_1}\int_{T_1}\left\{\frac{1}{2\,T_{sw}}\int_{T_{sw}}\left(\Delta I_K^{\alpha^2}(t) + \Delta I_K^{\beta^2}(t)\right)dt\right\}d\phi \\ &= \frac{1}{2\pi}\int_{2\pi}\Delta I^2_{rms(local)}(\phi), \end{aligned} \tag{4.33}$$

where, $\Delta I^2_{rms(local)}$, is the local (with respect to the switching cycle) per phase square RMS value of current ripples. This is called local RMS value of current ripples.

Based on equations (4.31), (4.32) and (4.33) the local and global RMS value of current ripples can be estimated for the SVPWM techniques.

4.5.1 Local Current Ripple RMS value

For the C-SVPWM technique, the switching sequence 01277210 is used in the voltage sector I of SVM. The local RMS value of current ripples is calculated from:

$$\Delta I^2_{rms(local)}(t) = \frac{1}{2}\left(\Delta I^{\alpha^2}_{rms(local)}(t) + \Delta I^{\beta^2}_{rms(local)}(t)\right). \qquad (4.34)$$

Taking the α-component for example, $\Delta I^{\alpha^2}_{rms(local)}(t)$ is given by:

$$\Delta I^{\alpha^2}_{rms,local}(t) = \frac{1}{T_{sw}}\left\{\sum_K \int_{t_{K,begin}}^{t_{K,end}} \Delta I^{\alpha^2}_K(t)\,dt\right\} \quad \text{for} \quad K = 0,1,2,7,7,2,1,0, \qquad (4.35)$$

where $t_{K,end}$ and $t_{K,begin}$ are the end and beginning times of switching the voltage vector $\vec{V}_K$. ΔI^{α}_K is given by (4.31) and (4.32) for the non-zero and zero voltage vectors respectively. Note, that due to the periodical essence of current waveform the value of current deviation at the beginning of each switching cycle equals this value at the end of that cycle. This value is assumed to be zero for simplicity of calculation. It is true in case of current control method described in chapter 6. For other cases the interval, in which the current deviation is calculated, can be shifted to the crossing point of the output current and fundamental current waveforms which leads to the same results. Therefore for the first voltage vector of the switching sequence, $\vec{V}_0$, ΔI^{α}_0 can be given by:

$$\Delta I^{\alpha}_0(t) = \frac{1}{L}\left(-\frac{m\,V_{dc}}{2}\cos(\phi)\right)(t), \qquad (4.36)$$

and for the next voltage vector, $\vec{V}_1$, considering $\Delta I^{\alpha}_1(t_{1,begin}) = \Delta I^{\alpha}_0(t_{0,end})$:

$$\begin{aligned}\Delta I^{\alpha}_1(t) &= \frac{1}{L}\left(\frac{2}{3}V_{dc} - \frac{m\,V_{dc}}{2}\cos(\phi)\right)(t - \frac{T_{0/7}}{2}) - \frac{m\,V_{dc}}{2\,L}\frac{T_{0/7}}{2}\cos(\phi)\\ &= \frac{V_{dc}}{L}\left[\left(\frac{2}{3} - \frac{m}{2}\cos(\phi)\right)t - \frac{T_{0/7}}{3}\right].\end{aligned} \qquad (4.37)$$

After deriving the current deviation equation of the other voltage vectors and introducing the square of these equations to (4.35), the following expression is found for the α-component of the local RMS value of current ripples:

$$\begin{aligned}\Delta I^{\alpha^2}_{\binom{rms,local}{c-svpwm}}(\phi) &= \Delta I^2_B\frac{m^2}{576}\left\{8\cos(\phi)\left(6\cos(\phi) - 9\,m\right) - 2\sqrt{3}\,m\sin^3(\phi)\right.\\ &\left.\left(9\,m\cos(\phi) - 8\right) + 27\,m^2\left(1 + \sin^2(\phi) - 2\sin^4(\phi)\right)\right\},\end{aligned} \qquad (4.38)$$

with

$$\Delta I_B = \frac{V_{dc}\, T_{sw}}{8\, L}. \tag{4.39}$$

In a same manner the β-component is obtained:

$$\begin{aligned} \Delta I^{\beta 2}_{\binom{rms,local}{c-svpwm}}(\phi) &= \Delta I_B^2 \frac{m^2}{576}\Big\{3\,\sin^2(\phi)\,\Big[16 + 27\,m^2 - 6\,m^2\sin^2(\phi) - \\ &\quad 2\sqrt{3}\,m\,\sin(\phi)\Big(9\,m\,\cos(\phi) - 8\Big)\Big]\Big\}. \end{aligned} \tag{4.40}$$

From (4.34) the local RMS value of current ripples is given by:

$$\begin{aligned} \Delta i^2_{\binom{rms,local}{c-svpwm}}(\phi) &= \frac{m^2}{576}\frac{1}{2}\Big\{48 + 9\,m\,\Big(3\,m - 8\,\cos(\phi)\Big) - 32\sqrt{3}\,m\,\sin^3(\phi) \\ &\quad \Big(2\,m\,\cos(\phi) + 1\Big) + 36\,m^2\,\sin^2(\phi)\,\Big(1 + 2\,\cos^2(\phi)\Big)\Big\}, \end{aligned} \tag{4.41}$$

which is normalized to ΔI_B. ΔI_B, which is the per phase square RMS value of the current deviation caused by $V_{dc}/2$ during one sampling cycle $T_S = T_{sw}/2$, is selected as the base value for current deviation. This allows to express equation (4.41) independent from V_{dc}, T_{sw} and L.

In a similar way the normalized per phase local RMS value of current ripples for the discontinuous SVPWM techniques can be calculated. The calculations are performed by the program Maple V algebra [45] and [46]. The results in form of three-dimensional surfaces are illustrated for voltage sector I in Fig. 4.25 for all SVPWM techniques.

As can be seen from this figure, the RMS value of current ripples is larger in the middle of the voltage sectors due to equality of T_K and T_{K+1}. It becomes small at low modulation indices because T_K and T_{K+1} become small, too. The RMS value of current ripples of ML-SVPWM at $\varphi = \pm 30^o$ are complementary to each other. The switching patterns of ML-SVPWM at $\varphi = \pm 30^o$ (127721 and 210012 for the voltage sector I) are the substances of switching

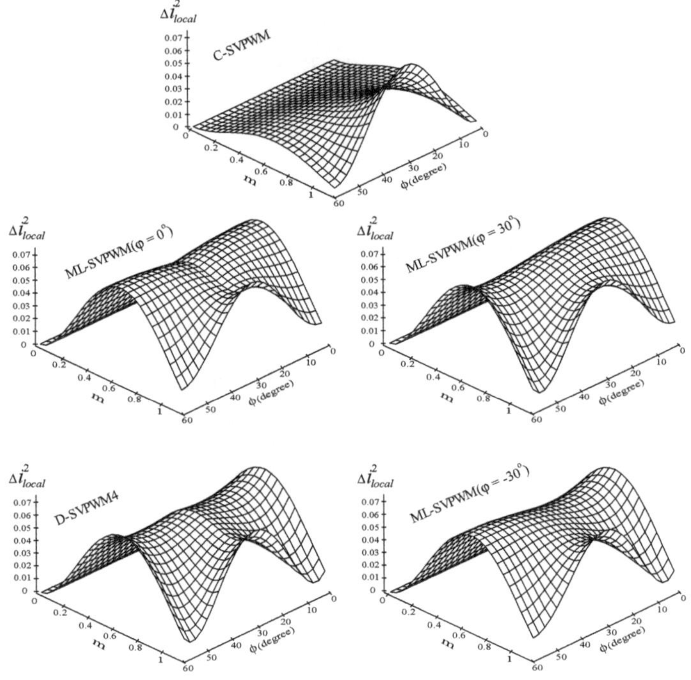

Fig. 4.25: Normalized per phase local square RMS value of current ripples versus modulation index m and ϕ for SVPWM techniques.

pattern of other discontinuous techniques under study. The symmetry of surfaces related to ML-SVPWM($\varphi = 0$) and D-SVPWM4 respect to the middle of sector is because of this reason. Considering the surface related to D-SVPWM4 for example, D-SVPWM4 uses the switching sequences 210012 and 127721 during the first and second half of the voltage sector I respectively, the surface (local RMS value of current ripples) related to D-SVPWM4 is the same as that of ML-SVPWM($\varphi = -30$) for the first half (0°-30^0) and as that of ML-SVPWM($\varphi = 30$) for the second half (30°-60^0) of this sector. The surface corresponding to the local RMS value of current ripples for the D-SVPWM5 is not shown because it is a combination of these substances, too. Comparing the switching pattern of D-SVPWM5 with those of ML-SVPWM($\varphi = 0$) and

D-SVPWM4, it can be easily concluded that the surface of D-SVPWM5 is similar to that of ML-SVPWM($\varphi = 0$) for the odd number of voltage sectors and similar to that of D-SVPWM4 for the even number of voltage sectors. The surface corresponding to the local RMS value of current ripples for the ST-SVPWM technique is also the same as that of ML-SVPWM technique at any power factor.

4.5.2 Global Current Ripple RMS value and Comparison

From the practical point of view, switching losses are more important than the switching frequency for the inverter devices. Therefore for the comparison of the different SVPWM strategies concerning the current waveform quality, equal switching losses is a preferable base. The switching losses are a dominative parameter in determination of the admissible switching frequency. Comparing the switching losses of different D-SVPWM techniques with those of C-SVPWM technique in section 4.4, we have shown that a reduction of 60%, 45% and 35% in switching energy losses and also in energy losses of the auxiliary circuit using ML-SVPWM, D-SVPWM5 and D-SVPWM4 techniques respectively, could be achieved (see Fig. 4.23 and 4.24). Since switching losses and switching frequency are proportionally related to each other, the switching frequency can be increased by the factor $(1/R_i)$ for D-SVPWM strategies where R_i is the switching energy loss ratio of discontinuous modulation techniques given by (4.25), (4.27) and (4.28). Using ML-SVPWM, D-SVPWM5 and D-SVPWM4 techniques, respectively a 2.5, 1.8 and 1.5 fold increase in switching frequency is possible in case of $\varphi = 0$.

According to (4.38) and (4.39) the local RMS value of current ripples decreases proportionally with the increase of the switching frequency. Therefore in order to generate the same switching energy losses for all techniques, their normalized local RMS value is divided by the respective factor, namely $(1/R_i)^2$, when a comparison among the global values is concerned.

According to (4.33), averaging of the normalized local RMS value of current ripples over the fundamental cycle results the global value. This value has been calculated for all purposed SVPWM strategies using Maple V. The respective normalized global values, which only depend on the modulation index m, are given as follows:

$$\Delta i^2_{\binom{rms,global}{c-svpwm}}(m) = \frac{m^2}{6}\left\{1 - \frac{8\,m}{\sqrt{3}\,\pi}\,\frac{9\,m^2}{8}\left(1 - \frac{3\sqrt{3}}{4\,\pi}\right)\right\}, \tag{4.42}$$

$$\Delta i^2_{\binom{rms,global}{ml-svpwm}}(m,\,\varphi) = \frac{m^2\,R^2_{ml-svpwm}(\varphi)}{6}\left\{4 - \frac{4\,m}{\sqrt{3}\,\pi}\Big[2 - 3\sqrt{3}\,\cos(\varphi)\right.$$
$$\left.\left(\cos^2(\varphi) - \frac{9}{4}\right)\Big] + \frac{9m^2}{4}\Big[1 - \frac{2\sqrt{3}\cos^2(\varphi)}{\pi}\left(\cos^2(\varphi) - \frac{3}{2}\right)\Big]\right\}, \tag{4.43}$$

$$\Delta i^2_{\binom{rms,global}{st-svpwm}}(m,\,\varphi) = 4\,\Delta i^2_{\binom{rms,global}{ml-svpwm}}(m,\,\varphi), \tag{4.44}$$

$$\Delta i^2_{\binom{rms,global}{d-svpwm4}}(m) = \frac{m^2 R^2_{d-svpwm4}}{6}\left\{4 - \frac{m}{\sqrt{3}\pi}\left(62 - 15\sqrt{3}\right)\right.$$
$$\left. + \frac{9m^2}{8}\left(2 + \frac{\sqrt{3}}{\pi}\right)\right\}, \tag{4.45}$$

$$\Delta i^2_{\binom{rms,global}{d-svpwm5}}(m,\varphi) = \frac{m^2 R^2_{d-svpwm5}(\varphi)}{6}\left\{4 - \frac{35m}{\sqrt{3}\pi} + \frac{9m^2}{8}\left(2 + \frac{3\sqrt{3}}{4\pi}\right)\right\}, \tag{4.46}$$

where $R_{ml-svpwm}$, $R_{d-svpwm4}$ and $R_{d-svpwm5}$ are given by (4.25), (4.27) and (4.28) respectively.

The dependence of descriptions given above on the modulation index is shown in Fig. 4.26(a). It is obvious that under the circumstance of the same switching losses the ML-SVPWM strategy has the lowest global RMS value of current ripples for the whole modulation index range, if $-\frac{\pi}{6} < \varphi < \frac{\pi}{6}$. Using this technique a 2.5 fold increase in the switching frequency can be achieved resulting in the best waveform quality for the output current. This is advantageous from the point of view of volume/costs of AC/EMI filters and of the series inductances in rectifiers and of view of the losses caused by high order harmonics of output current in AC machine loads of inverters.

The global RMS value of current ripples of ML-SVPWM technique drastically increases for $\varphi < -\frac{\pi}{6}$ & $\varphi > \frac{\pi}{6}$. At $\varphi = \frac{\pi}{3}$ this value becomes even greater than that for C-SVPWM despite a two fold higher switching frequency. Figure 4.26(b) shows the variation of the global RMS value of current ripples versus the current

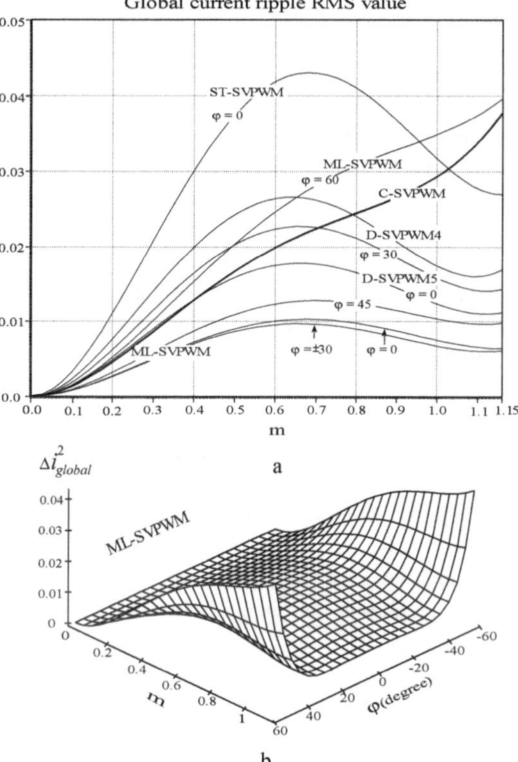

Fig. 4.26: Normalized per phase global square RMS value of current ripples a) versus modulation index m for all SVPWM techniques and b) versus modulation index m and current phase angle φ for ML-SVPWM technique.

phase angle φ for the ML-SVPWM technique.

If the comparison is made under the same switching frequency, then all waveforms of the global RMS value of current ripples related to the discontinuous strategies are positioned above the waveform of the continuous strategy. In this case the advantage of lower switching losses using the discontinuous modulation methods can be utilized for increasing the power density of converter.

4.6 Summary of Chapter

The commutated pole concept of AC-Side Soft Commutated PWM Converters (AC-Side Half/Quarter Wave Auxiliary Resonant Commutated Pole circuits) do not restrict the modulation technique used for switching the converter switches. Thus, the Space Vector PWM (SVPWM) can be optimized in such a way that the switching losses of converter including the losses of auxiliary circuit are minimized.

This chapter investigates the continuous and different discontinuous space vector PWM in detail. A generalized discontinuous Minimum-Loss Space Vector PWM (ML-SVPWM) technique is proposed for minimizing the switching losses for a wide range of variation of current phase angle φ. For $\varphi \in [-\frac{\pi}{6}, \frac{\pi}{6}]$ the technique avoids switching within the 60° intervals located in vicinity of current peaks. In this way a reduction of 60% in switching losses of main circuit and also in the losses of auxiliary circuit can be achieved. For the ranges $\varphi \in [-\frac{\pi}{3}, -\frac{\pi}{6}]$ and $\varphi \in [\frac{\pi}{6}, \frac{\pi}{3}]$ switching at the phase with the second absolutely largest current magnitude is avoided. Reduction of switching losses in these ranges depends on the current phase angle. At $\varphi = \pm\frac{\pi}{3}$ a reduction of 45% can be expected. In order to investigate the influence of the different space vector PWM technique on the amplitude of higher order harmonics of phase current, the RMS value of current ripples for different modulation techniques is calculated. The results are compared with each other under the same switching losses (different switching frequency). It is concluded that the RMS value of current ripples for the Minimum-Loss Space Vector PWM (ML-SVPWM) in the range $\varphi \in [-\frac{\pi}{6}, \frac{\pi}{6}]$ is lower than that of other PWM techniques. If the comparison is made under the same switching frequency, the RMS value of current ripples of all discontinuous space vector PWM techniques would be higher than that of continuous modulation. In this case one can benefit from the advantage of lower switching losses of discontinuous space vector modulation techniques for increasing the power density of converter.

This chapter also introduces a new discontinuous switching pattern which changes the zero voltage vectors every 30° interval instead of every 60° as in the case of other well-known discontinuous switching patterns. This reduces the temperature excursions on the power devices which create the zero voltage vector within the mentioned intervals and therefore improves the reliability.

5. Influence of ZVS Durations on Space Vector Modulation

In this chapter, a new area of study is opened. The discussion of this chapter draws the influence of different stages of voltage transitions on the space vector modulation into consideration and reveals the restrictions which due to the necessary duration of these stages come into consideration. These restrictions mostly result in a not full utilization of DC bus voltage or in a reduction of linearity range of the modulation. The full utilization of the DC bus voltage is extremely important to obtain the voltage control margin even under increased source voltage [55]. Reduction of the DC bus utilization due to transition durations, effect of the necessary duration of voltage vectors on the availability of the entire voltage hexagon area of space vector modulation and on the linearity range of modulation index are the further subjects to be discussed in this chapter.

5.1 Effect of Voltage Transition Durations on the DC Bus Utilization

During the resonant state of an ACA and during the self-commutations (switch-to-diode current commutations) node voltages V_a, V_b and V_c do not generate the voltage vector required in the space vector modulation. This affects the synthesis of the reference voltage vector $\vec{V}_{ref}$ resulting in the modulation nonlinearity and reduces the DC bus utilization. At high switching frequencies these effects become more visible.

Fig. 5.1 shows an example for the α-component volt-second areas resulting from the synthesis of $\vec{V}_{ref}$ by hard switching circuits, $A^{\alpha}_{VS(SVM-HS)}$, soft switching circuits, $A^{\alpha}_{VS(SVM-SS)}$, and the reference one $A^{\alpha}_{VS(ref)}$. Since the space vector modulation is based on the equilibrium of volt-second area for both α

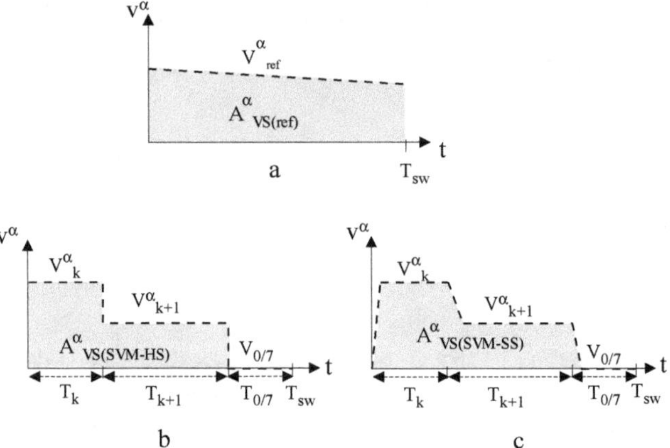

Fig. 5.1: α-component of the volt-second area vector: a) desired, b) synthesized by a hard switching converter and c) synthesized by a soft switching converter

and β components, a volt-second area vector over the switching cycle T_{sw} can be defined which in case of a hard switching converter, $\vec{A}_{VS(SVM-HS)}$, is always equal to the desired reference volt-second area vector, $\vec{A}_{VS(ref)}$. But in case of soft switching converters, these volt-second area vectors are not equal, due to the mostly larger commutation durations causing larger volt-second losses.

In this section the effect of the transition durations on the modulated AC voltage is discussed and a method for compensating this effect is suggested. The discussion is established on the switching sequence $S_{1(pnn)} - S_{7(ppp)} - S_{2(ppn)} - S_{1(pnn)}$ used also in Chapter 2 for describing the operation of ASSC converters. The phase currents I_a, I_b and I_c are assumed to be large enough and positive, negative and negative respectively. Both the discussion and the compensation method can be evidently applied to any other switching sequence and load condition in a similar way.

With the assumed phase current directions, the transition between the switching states $S_{1(pnn)}$ and $S_{7(ppp)}$ requires the actuation of the auxiliary circuit, while the others are self-commutations and take place by opening the main switches, one at the time.

5.1.1 Effect of the Resonant State Duration

The transition between the switching states $S_{1(pnn)}$ and $S_{7(ppp)}$ producing the voltage vectors $\vec{V}_{1(pnn)}$ and $\vec{V}_{7(ppp)}$ takes place during the resonant state. In this state the node voltage V_a is clamped to the positive DC rail voltage, while the other node voltages V_b and V_c increase in resonant fashion from the negative DC rail voltage to the positive DC rail voltage. In the three-phase coordination system the vector of the node voltages during the resonant state, $\vec{V}_{res,abc}$, for the ACS-HW-ARCP circuit ($I_{boost} = 0$) is described by:

$$\vec{V}_{res,abc}(t) = \begin{bmatrix} \frac{V_{dc}}{2} \\ -\frac{V_{dc}}{2}\cos(\omega_r t) \\ -\frac{V_{dc}}{2}\cos(\omega_r t) \end{bmatrix}. \tag{5.1}$$

Transforming this vector into the stationary $\alpha\beta$ reference frame yields:

$$\vec{V}_{res,\alpha\beta}(t) = \left(\frac{1+\cos(\omega_r t)}{2}\right)\frac{2}{3}V_{dc}\begin{bmatrix} 1 \\ 0 \end{bmatrix} = \left(\frac{1+\cos(\omega_r t)}{2}\right)\vec{V}_{1(pnn)}. \tag{5.2}$$

The above expression shows that a transition voltage vector in the direction of $\vec{V}_{1(pnn)}$ with a variable magnitude is generated during the resonant state. The resulting volt-second area vector $\vec{A}_{VS(res)}$ can be calculated as follows:

$$\begin{aligned} \vec{A}_{VS(res)} &= \int_0^{\Delta t_{res}} \vec{V}_{res,\alpha\beta}(t)\,dt = \left(\frac{\Delta t_{res}}{2} + \frac{1}{2\,\omega_r}\sin(\omega_r\,\Delta t_{res})\right)\frac{2}{3}V_{dc}\begin{bmatrix} 1 \\ 0 \end{bmatrix} \\ &\approx \left(\frac{\Delta t_{res}}{2}\right)\frac{2}{3}V_{dc}\begin{bmatrix} 1 \\ 0 \end{bmatrix} = \frac{\Delta t_{res}}{2}\vec{V}_{1(pnn)}. \end{aligned} \tag{5.3}$$

This means, that the volt-second area vector corresponding to $\vec{V}_{1(pnn)}$ is increased by $\vec{A}_{VS(res)}$. Thus, for compensating its effect, the duty cycle of $\vec{V}_{1(pnn)}$, T_1, should be reduced by $\Delta t_{res}/2$ and the duty cycle of zero-vector $\vec{V}_{7(ppp)}$, T_7, should be increased by the same amount. This compensation method, i.e. equal distribution of transition duration between the duty cycle of voltage vectors preceding and following the transition, establishes again an equilibrium between the resulting volt-second areas.

5.1.2 Effect of the self-Commutation Duration

The transition between the switching states $S_{7(ppp)}$ and $S_{2(ppn)}$ producing the voltage vectors $\vec{V}_{7(ppp)}$ and $\vec{V}_{2(ppn)}$ is an example of the current self-commutation, which takes place by opening the switch S_{cp}. The node voltage vector during this commutation, $\vec{V}_{sc,abc}$, can be described by:

$$\vec{V}_{sc,abc}(t) = \begin{bmatrix} \frac{V_{dc}}{2} \\ \frac{V_{dc}}{2} \\ \frac{V_{dc}}{2}\left(1 - \frac{2t}{\Delta t_{sc}}\right) \end{bmatrix}. \tag{5.4}$$

Applying the $\alpha\beta$ transformation matrix to $\vec{V}_{sc,abc}(t)$ results in:

$$\vec{V}_{sc,\alpha\beta}(t) = \frac{t}{\Delta t_{sc}} \frac{2}{3} V_{dc} \begin{bmatrix} \frac{1}{2} \\ \frac{\sqrt{3}}{2} \end{bmatrix} = \frac{t}{\Delta t_{sc}} \vec{V}_{2(ppn)}. \tag{5.5}$$

This indicates that during the transition between a zero voltage vector and a non-zero voltage vector a transition voltage vector with the time variable magnitude in the direction of the non-zero vector is produced. The volt-second area vector $\vec{A}_{VS(sc)}$ resulting from this vector is obtained from:

$$\vec{A}_{VS(sc)} = \int_0^{\Delta t_{sc}} \vec{V}_{sc,\alpha\beta}(t)\, dt = \frac{\Delta t_{sc}}{2} \vec{V}_{2(ppn)}. \tag{5.6}$$

It reveals that the volt-second area vector related to $\vec{V}_{2(ppn)}$ is increased by $\vec{A}_{VS(sc)}$. Thus, according to the proposed compensation method, the duration of the current self-commutation, Δt_{sc}, should be divided equally between the duty-cycle of $\vec{V}_{7(ppp)}$ and $\vec{V}_{2(ppn)}$ to compensate this effect.

The transition between the switching states $S_{2(ppn)}$ and $S_{1(pnn)}$ producing $\vec{V}_{2(ppn)}$ and $\vec{V}_{1(pnn)}$ is another example of the current self-commutation, which takes place by opening the switch S_{bp}. The transition voltage vector in the three phase coordination system and in the α, β reference frame is given by following equations:

$$\vec{V}_{sc,abc}(t) = \begin{bmatrix} \dfrac{V_{dc}}{2} \\ \dfrac{V_{dc}}{2}\left(1 - \dfrac{2t}{\Delta t_{sc}}\right) \\ -\dfrac{V_{dc}}{2} \end{bmatrix}, \tag{5.7}$$

$$\vec{V}_{sc,\alpha\beta}(t) = \frac{2}{3}V_{dc} \begin{bmatrix} \dfrac{1}{2}\left(1 + \dfrac{t}{\Delta t_{sc}}\right) \\ \dfrac{\sqrt{3}}{2}\left(1 - \dfrac{t}{\Delta t_{sc}}\right) \end{bmatrix}. \tag{5.8}$$

The resulting volt-second area vector is calculated as follows:

$$\begin{aligned} \vec{A}_{VS(sc)} &= \int_0^{\Delta t_{sc}} \vec{V}_{sc,\alpha\beta}(t)\,dt = \left(\frac{\Delta t_{sc}}{2}\right)\frac{2}{3}V_{dc}\begin{bmatrix} 1 + \dfrac{1}{2} \\ 0 + \dfrac{\sqrt{3}}{2} \end{bmatrix} \\ &= \frac{\Delta t_{sc}}{2}\vec{V}_{1(pnn)} + \frac{\Delta t_{sc}}{2}\vec{V}_{2(ppn)}. \end{aligned} \tag{5.9}$$

This proves again the proposed method, i. e. the duration of the transition has to be divided equally between the voltage vectors preceding and following it, namely $\vec{V}_{2(ppn)}$ and $\vec{V}_{1(pnn)}$. In other words the duration of $\vec{V}_{2(ppn)}$ preceding the transition should be reduced by $\Delta t_{sc}/2$ while the duration of $\vec{V}_{1(pnn)}$ following the transition should be increased by the same amount.

To show the effect of the transition duration on the reduction of the DC bus utilization and the effectiveness of the compensation method, simulations with and without using the compensation method are performed for the pole commutating circuits under the same DC bus voltage, the same auxiliary circuit characteristics and at the same modulation index. The AC voltages generated with different switching frequencies are analysed using the Fourier analysis. The amplitudes of the fundamental component of the modulated voltages are compared with each other as a measure for DC bus utilization. The results are illustrated in Fig. 5.2. The reduction of the DC bus utilization can be seen from the reduction of the fundamental component amplitude of the generated AC voltage. Compared to the hard switching circuit, respectively a 1.1% and 2.3% reduction in the DC bus utilization can be seen for the ARC-QW-ARCP and ACS-HW-ARCP circuits at 30kHz switching frequency in Fig. 5.2(a). This reduction becomes more

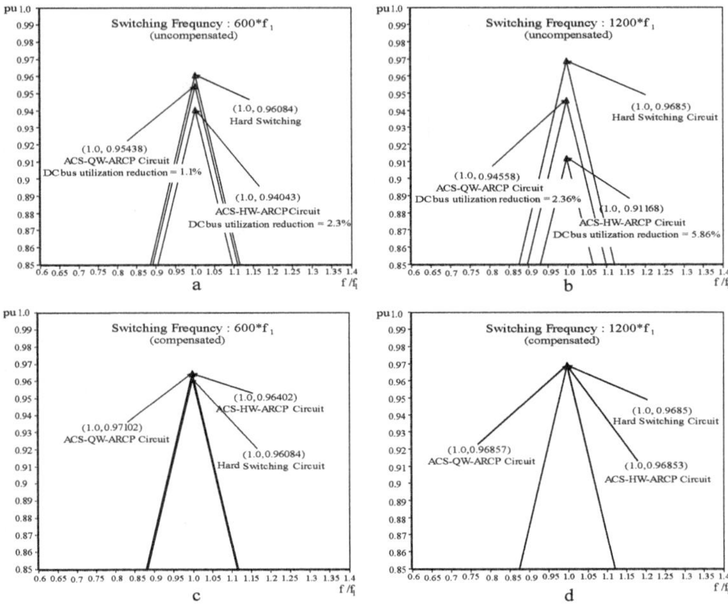

Fig. 5.2: a and b) The effect of the transition durations on the DC bus utilisation reduction and c and d) The effectiveness of the compensation method

considerable at higher switching frequencies, so that it reaches a value as much as 2.36% and 5.86% for the ARC-QW-ARCP and ARC-HW-ARCP circuits at 60kHz (Fig. 5.2 (b)). Reduction of the DC bus utilization means that a higher DC voltage is necessary for generating the required AC voltages resulting in an increase of losses. Using the proposed compensation method, the reduction of DC bus utilization can effectively be compensated. It is confirmed in Fig. 5.2(c and d). Since the computation of transition duration and of the duty-cycle of voltage vectors can be performed altogether in the computing unit of a DSP-based controller, implementation of this method is easy and has no additional costs.

5.2 Effect of the Charging and Discharging Durations on the Linearity Range of Modulation

The influence of duration of the transient state of a ZVS commutation on SVM is discussed in the last section. This section targets the influence of duration of other states of a ZVS commutation on SVM and introduces some expressions, which simply describe these influences. The duration of polarity changing of phase voltages will not be considered in order to separate the discussion of this section from that of last section.

The duration of charging and discharging states including turn-on/off delay time of devices limits the minimum duty cycle of the voltage vectors in SVPWM. Since uncompleted ZVS processes result in increase of the switching losses and unnecessary switching actions, soft current commutation in each phase should be completely finished before a new voltage vector, which requires switching at the same phase, can be selected. Considering this fact, the critical vectors with extremely short duty cycle should be eliminated to avoid uncompleted ZVS processes. In this section it will be shown that at high modulation indices zero voltage vectors become critical and that the elimination of these vectors results in reduction of the linearity range of modulation. The non-zero vectors become also critical in high frequency applications. These effects will be stated precisely in this section for two basic modulation strategies, namely:

1. `Continuous or Three Phase Modulation` (CM),
2. `Discontinuous or Two Phase Modulation` (DM).

Assuming the reference voltage vector, $\vec{V}_{ref}$, in sector I of the voltage hexagon and the vector of phase currents, $\vec{I}$, in the second half of sector a of the current hexagon ($I_a > 0 > I_b > I_c$) as shown in Fig. 5.3, the switching sequences:

1. $S_{0(nnn)} - S_{1(pnn)} - S_{2(ppn)} - S_{7(ppp)} - S_{7(ppp)} - S_{2(ppn)} - S_{1(pnn)} - S_{0(nnn)}$ and,
2. $S_{1(pnn)} - S_{2(ppn)} - S_{7(ppp)} - S_{7(ppp)} - S_{2(ppn)} - S_{1(pnn)}$

are used in the continuous and discontinuous modulation strategies respectively. Because of the symmetry of modulation and of three-phase quantities the same results can be concluded for other sectors of voltage and current hexagons, if the proper switching sequence is considered (see chapter 4).

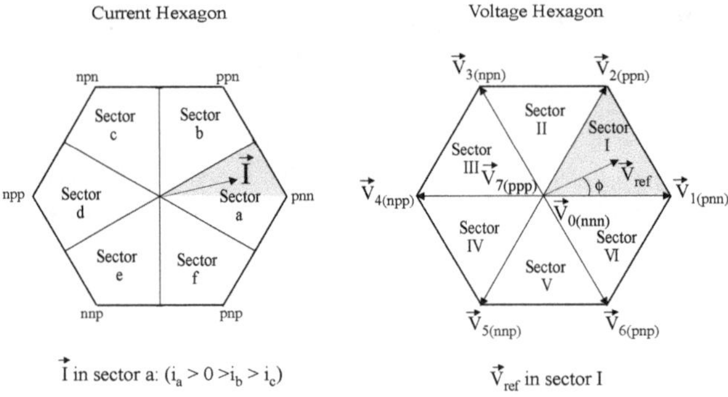

Fig. 5.3: Voltage and current hexagons

The duration of a current self-commutation (switch-to-diode current commutation) is assumed to be shorter than the total duration of a diode-to-switch ZVS current commutation for the same phase current magnitude. In the self-commutation of large phase currents the phase current charges and discharges the node capacitances fast enough. In case of small phase currents the auxiliary circuit is activated to accelerate the current commutation. Because of the shorter charging and discharging durations in this case, the total commutation duration of this commutation is also shorter than that of a diode-to-switch ZVS current commutation. Therefore, the necessary time duration of a diode-to-switch ZVS commutation will be considered as a determinative duration in the following discussion.

The discussion will only cover the commutated pole circuits. The other circuits are excluded because they can only operate with the suboptimal modulation techniques in which turn-on instants of main switches should be synchronized in ZVS commutations.

5.2.1 Maximum Available Continuous Linear Modulation Index

Table 5.1 shows the polarity of the node voltages along with the kind of commutation at the time of polarity changing for the continuous modulation.

Continuous (Three Phase) Modulation														
Seq.	S_0	$\rightarrow$	S_1	$\rightarrow$	S_2	$\rightarrow$	S_7	S_7	$\rightarrow$	S_2	$\rightarrow$	S_1	$\rightarrow$	S_0
Ph.	**Phase Polarity and Kind of Transition**													
a	n	StD	p		p		p	p		p		p	DtS	n
b	n		n	DtS	p		p	p		p	StD	n		n
c	n		n		n	DtS	p	p	StD	n		n		n
Ph.	**Kind of Commutation and Auxiliary Branch States**													
a	$\rightarrow$	SC									$\leftarrow$	CS	TS	DS
b		$\leftarrow$	CS	TS	DS	$\rightarrow$					SC			
c				$\leftarrow$	CS	TS	DS	$\rightarrow$	SC					

Table 5.1: Phase polarity (phase is clamped to p: positive or n: negative DC rail) and states of auxiliary branch of each phase for continuous modulation considering $\vec{V}_{ref}$ in sector I and $I_a > 0 > I_b > I_c$ (CS: Charging State, TS: Transient State, DS: Discharging State and SC: Self-Commutation)

The abbreviations DtS and StD refer to the diode-to-switch and switch-to-diode current commutations, respectively. The different states of the auxiliary branches, Charging State (CS), Transient State (TS) and Discharging State (DS), and the Self-Commutation (SC) are also shown in this table. These informations are shown for phases a and c in Fig. 5.4.

It can be drawn from the figure and table that in order to achieve a complete diode-to-switch ZVS current commutation for the transition $S_1 \rightarrow S_0$ at phase a, the duty cycles of S_1 and S_0 *should not be* absolutely longer than the necessary times for the charging and discharging states of the transition respectively. In other words the charging state of the auxiliary branch at phase a can begin any time while this phase is clamped to the positive DC rail, i. e. immediately after switching on the switching state S_1 in the preceding sampling cycle. Analogous the discharging state can end any time while phase a is clamped to the negative DC rail, i. e. until the end of the duration of S_0 in the next coming cycle. Paying more attention to the table, it is also visible that the critical durations are the discharging durations of the DtS-ZVS commutations at phases a and c in transitions $S_1 \rightarrow S_0$ and $S_2 \rightarrow S_7$ respectively. The discharging state of these

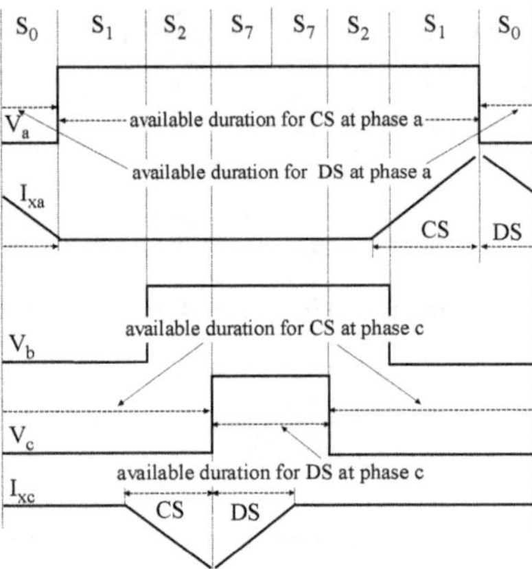

Fig. 5.4: Phase voltages V_a, V_b and V_c and auxiliary currents I_{xa} and I_{xc} during one switching cycle for continuous modulation.

transitions has to be finished before the phase voltages V_a and V_c change their polarity. So particularly the switching states S_7 and S_0, corresponding to the zero voltage vectors, can become critical in high modulation indices when their duty cycle is very short. From this point of view the condition of switching the zero voltage vectors can be given as:

$$T_{0/7} > \Delta t_{dis}. \tag{5.10}$$

Substituting $T_{0/7}$ from Eq. (4.3) follows:

$$T_S \left\{ 1 - \frac{\sqrt{3}}{2} m_{cm} \left[\sin\left(\frac{\pi}{3} + \phi \right) \right] \right\} > \Delta t_{dis}, \tag{5.11}$$

where $T_S = 1/f_S$ is the sampling cycle and m_{cm} is the modulation index for the continuous modulation technique[1)] defined by (4.4).

The zero voltage vectors have their shortest duration at $\phi = 30^o$. Substituting it in (5.11) yields:

[1)] To differentiate the modulation indices of the different modulation techniques from each other, m_{cm} with the index *cm* and m_{dm} with the index *dm* are defined for the continuous and discontinuous modulation techniques respectively.

$$T_S\left(1-\frac{\sqrt{3}}{2}\,m_{cm}\right) > \Delta t_{dis}. \tag{5.12}$$

This expression indicates that the maximum linear value of the modulation index, $m_{max} = 1.15$, cannot be achieved for the ASSC PWM converters because Δt_{dis} is always greater than zero. For a certain sampling frequency and phase current amplitude, the maximum linear continuous modulation index for both the ACS-HW-ARCP and ACS-QW-ARCP circuits, $m_{cm,HW,max}$ and $m_{cm,QW,max}$, can be calculated by substituting Δt_{dis} from (3.57) and (3.78) in (5.12). It follows:

$$\begin{cases} m_{cm,HW,max} &= \dfrac{2}{\sqrt{3}}\left[1 - f_S\left(\Delta t + \dfrac{2\,Z_r\,I_a(\phi=30^o)}{\omega_r\,V_{dc}}\right)\right], \\ m_{cm,QW,max} &= \dfrac{2}{\sqrt{3}}\left[1 - f_S\left(\dfrac{Z_r\,I_a(\phi=30^o)+V_{dc}}{\omega_r\,V_{dc}}\right)\right], \end{cases} \tag{5.13}$$

where $I_a(\phi=30^o) = \hat{I}\cos(30^o-\varphi)$. The angle φ is defined in Fig. 4.13(b). Selecting $V_B = \dfrac{V_{dc}}{2}$, $I_B = \hat{I}$ at full load condition and $f_B = f_1$ as the base values, resulting $Z_B = \dfrac{V_B}{I_B} = \dfrac{V_{dc}}{2\hat{I}}$, the above expressions can be given in the normalized form as follows:

$$\begin{cases} m_{cm,HW,max} &= \dfrac{2}{\sqrt{3}}\left[1 - \dfrac{f_S}{f_1}\left(\Delta t\,f_1 + \dfrac{z_r\,\hat{\imath}\,\cos(30^o-\varphi)}{(\omega_r/f_1)}\right)\right], \\ m_{cm,QW,max} &= \dfrac{2}{\sqrt{3}}\left[1 - \dfrac{f_S}{f_1}\left(\dfrac{z_r\,\hat{\imath}\,\cos(30^o-\varphi)+2}{2\,(\omega_r/f_1)}\right)\right], \end{cases} \tag{5.14}$$

which depend only on the normalized parameters of the auxiliary resonant circuit, z_r and ω_r, sampling frequency ratio, f_S/f_1, normalized amplitude, $\hat{\imath}$, and phase angle φ of output current. Fig. 5.5 shows the variation of the maximum available continuous modulation index, $m_{cm,max}$, versus these parameters for both circuits. The duration of discharging state in diode-to-switch ZVS current commutations determined by the auxiliary inductor(s), L_x, and DC link voltage strongly limits the available maximum linear continuous modulation index. This can be seen from Fig. 5.5(b). With a certain snubber capacitance C_r paralleled to main switches, increase of L_x resulting in longer discharging duration decreases ω_r and increases z_r. For $[(\omega_r/f_1) < 5\times10^4]$ further decrease of ω_r leads to

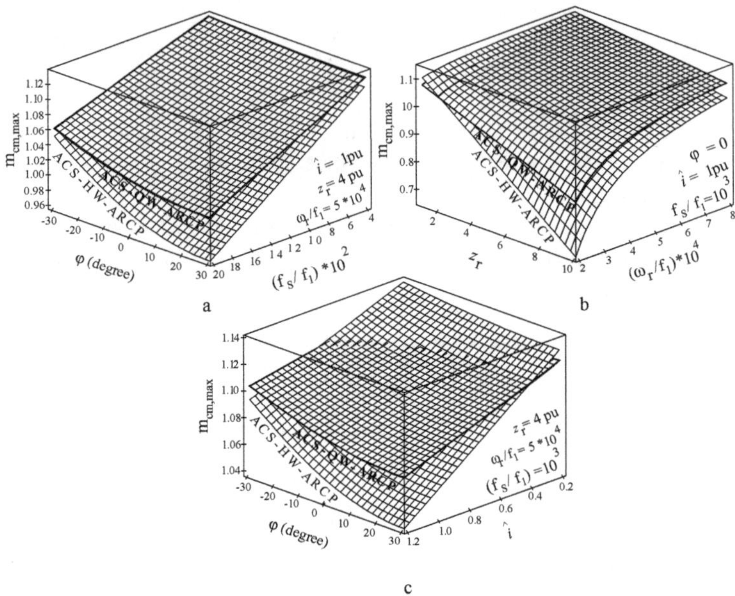

Fig. 5.5: $m_{cm,max}$ for both commutated pole concepts a) versus the auxiliary circuit characteristics (z_r and ω_r), b) versus f_S and φ and c) versus $\hat{i}$ and φ.

strong reduction of the available maximum linear continuous modulation index. The achievable linear maximum modulation index is also reduced by increasing the sampling frequency. This means for high power high frequency converters the linearity range of the modulation is considerably reduced because the zero voltage vectors can become critical even in modulation indices lower than one.

Because of full V_{dc} across the auxiliary inductors of ARC-QW-ARCP circuit during the discharging state, the linearity range of modulation in this circuit is wider than that of ARC-HW-ARCP circuit except at light loads. Since the auxiliary currents at the beginning of discharging state are as large as their maximum in the ARC-QW-ARCP circuit and almost as large as the phase currents in the ARC-HW-ARCP circuit, the discharging state of ARC-QW-ARCP circuit becomes longer than that of ARC-HW-ARCP circuit at light loads.

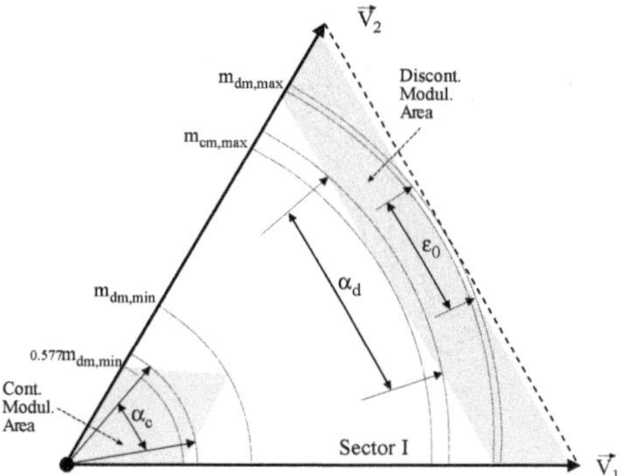

Fig. 5.6: Modulation areas, modulation changing angle α_d, continuous modulation utilization angle α_c and elimination angle ε_0. (α_c and ε_0 will be defined in continuation of this chapter.)

For $m > m_{cm,max}$ (above the surfaces shown in Fig. 5.5), $T_{0/7}$ can be divided no longer between $\vec{V}_{7(ppp)}$ and $\vec{V}_{0(nnn)}$. Thus one of the zero voltage vectors should be eliminated within the angle α_d in the middle of all voltage sectors. This results in changing the modulation strategy from the continuous to the discontinuous one as shown in Fig. 5.6. The angle α_d is called modulation changing angle.

The modulation changing angle α_d can be calculated by substituting Δt_{dis} from (3.57) and (3.78) in (5.11) resulting in the following equation (note the definition of α_d in Fig. 5.6):

$$\alpha_d = \pi - 2\ \arcsin\left(\frac{m_{cm,max}}{m}\right) \qquad \text{for} \quad m > m_{cm,max} \tag{5.15}$$

For $m > 2\, m_{cm,max}/\sqrt{3} = 1.15\, m_{cm,max}$, α_d become greater than 60° and the continuous modulation can no longer be used. Fig. 5.7 shows the variation of this angle versus m and $m_{cm,max}$. As it can be seen from this figure, the modulation changing angle, α_d, becomes considerably large at high switching frequencies where $m_{cm,max}$ is small. $m_{cm,max}$ is also dependent on the amplitude of the phase current (see Fig. 5.5(c)). This means that the continuous modulation can be

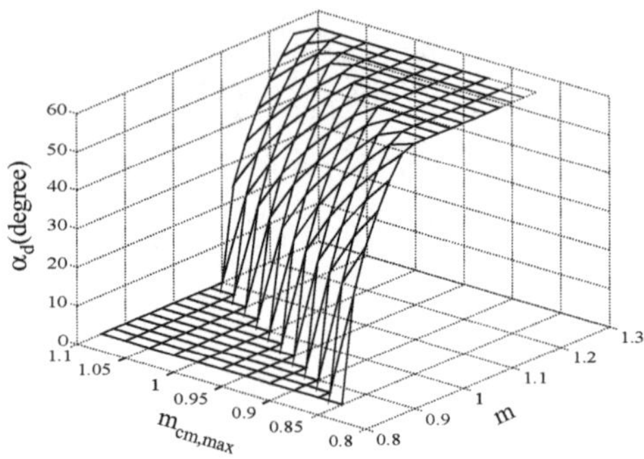

Fig. 5.7: Variation of angle α_d versus m and $m_{cm,max}$.

rarely used at high power, high switching frequency applications, especially when a high modulation index is also required.

5.2.2 Maximum Available Discontinuous Linear Modulation Index

The discontinuous modulation gives more time to the discharging state of ZVS by switching only one of the zero vectors and by placing the zero vectors of two sampling cycles side by side as well. This makes a higher linear modulation index possible.

Table 5.2 shows the same information as given in table 5.1 for the discontinuous modulation and Fig. 5.8 illustrates the phase voltages and the auxiliary current at phase c.

The duration of zero voltage vectors becomes shorter with the increase of the modulation index. As the modulation index further increases beyond $m_{cm,max}$, from a certain maximum discontinuous modulation index, $m_{dm,max}$, even the sum of the duty cycles of the two zero voltage vectors placed by each other is no longer enough for the discharging state of the auxiliary branch of phase c at the middle of the voltage sectors. They should be herein completely eliminated,

Discontinuous (Two Phase) Modulation										
Seq.	S_1	$\rightarrow$	S_2	$\rightarrow$	S_7	S_7	$\rightarrow$	S_2	$\rightarrow$	S_1
Ph.	**Phase Polarity and Kind of Transition**									
a	p		p		p	p		p		p
b	n	DtS	p		p	p		p	StD	n
c	n		n	DtS	p	p	StD	n		n
Ph.	**Kind of Commutation and Auxiliary Branch States**									
a										
b	CS	TS	DS	$\rightarrow$					SC	$\leftarrow$
c		$\leftarrow$	CS	TS	DS	$\rightarrow$	SC			

Table 5.2: Phase polarity (phase is clamped to p: positive or n: negative DC rail) and states of auxiliary branch of each phase for discontinuous modulation considering $\vec{V}_{ref}$ in sector I and $I_a > 0 > I_b > I_c$ (CS: Charging State, TS: Transient State, DS: Discharging State and SC: Self-Commutation)

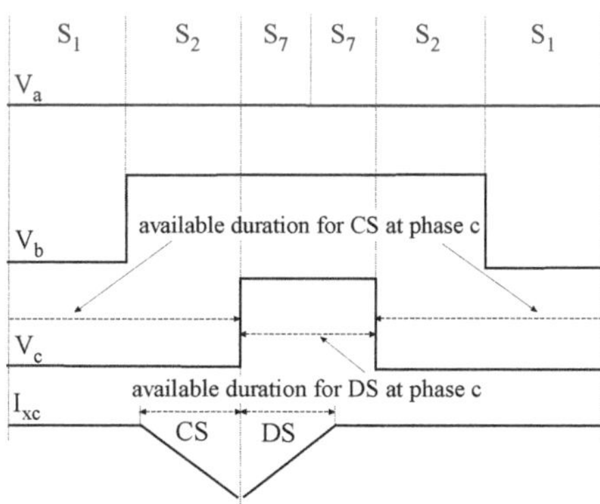

Fig. 5.8: Phase voltages V_a, V_b and V_c and the auxiliary current I_{xc} during one switching cycle for discontinuous modulation.

thereby the modulation enters the nonlinear area. This happens as soon as the following condition can be no longer fulfilled:

$$2\,T_{0/7} > \Delta t_{dis}. \tag{5.16}$$

Referring to equation (4.3) and substituting $\phi = 30^o$ and Δt_{dis} from (3.57) and (3.78) in (5.16) results in the maximum available linear discontinuous modulation index for the ACS-HW-ARCP and ACS-QW-ARCP circuits with $m_{dm,HW,max}$ and $m_{dm,QW,max}$, as follows:

$$\begin{cases} m_{dm,HW,max} &= \dfrac{1}{\sqrt{3}}\left[2 - f_S\left(\Delta t + \dfrac{2\,Z_r\,|I_c(\phi = 30^o)|}{\omega_r\,V_{dc}}\right)\right], \\ m_{dm,QW,max} &= \dfrac{1}{\sqrt{3}}\left[2 - f_S\left(\dfrac{Z_r\,|I_c(\phi = 30^o)| + V_{dc}}{\omega_r\,V_{dc}}\right)\right], \end{cases} \tag{5.17}$$

where $I_c(\phi = 30^o) = \hat{I}\cos(150^o - \varphi)$. Selecting the same base values used for normalizing (5.13), expressions above can be also normalized as follows:

$$\begin{cases} m_{dm,HW,max} &= \dfrac{1}{\sqrt{3}}\left[2 - \dfrac{f_S}{f_1}\left(\Delta t\,f_1 + \dfrac{z_r\,|\,\hat{i}\cos(150^o - \varphi)|}{(\omega_r/f_1)}\right)\right], \\ m_{dm,QW,max} &= \dfrac{1}{\sqrt{3}}\left[2 - \dfrac{f_S}{f_1}\left(\dfrac{z_r\,|\,\hat{i}\cos(150^o - \varphi)| + 2}{2\,(\omega_r/f_1)}\right)\right]. \end{cases} \tag{5.18}$$

For the modulation indices beyond $m_{dm,max}$ the zero vector has to be completely eliminated within the elimination angle ε_0 shown in Fig. 5.6. This angle is given by:

$$\varepsilon_0 = \pi - 2\,\arcsin\left(\frac{m_{dm,max}}{m}\right) \qquad \text{for } m > m_{dm,max}. \tag{5.19}$$

The Comparison of expressions given by (5.17) with those given by (5.14) results in:

$$\begin{cases} m_{dm,HW,max} &= \dfrac{1}{\sqrt{3}} + \dfrac{m_{cm,HW,max}}{2} + \dfrac{f_S}{f_1}\,\dfrac{z_r\,\hat{i}\,\cos(\varphi)}{(\omega_r/f_1)}, \\ m_{dm,QW,max} &= \dfrac{1}{\sqrt{3}} + \dfrac{m_{cm,QW,max}}{2} + \dfrac{f_S}{f_1}\,\dfrac{z_r\,\hat{i}\,\cos(\varphi)}{2\,(\omega_r/f_1)}, \end{cases} \tag{5.20}$$

which indicates the relationship of the maximum available modulation index of the discontinuous modulation with that of the continuous modulation for both commutated pole circuits. Fig. 5.9 illustrates this relationship for $\varphi = 0$. In

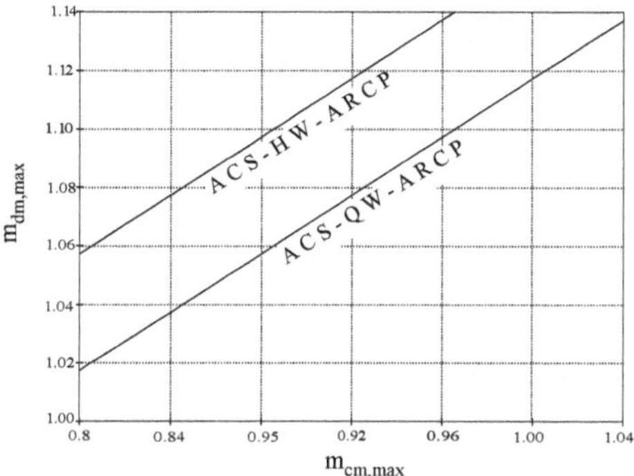

Fig. 5.9: $m_{dm,max}$ *versus* $m_{cm,max}$ $(z_r = 4pu, (\omega_r/f_1) = 5 \times 10^4, (f_S/f_1) = 10^3,$ $\hat{i} = 1pu$ *and* $\varphi = 0)$

this figure the advantage of the discontinuous modulation, i. e. availability of a higher maximum linear modulation index in the soft commutated converters, can be evidently seen. That means, with the discontinuous modulation DC link voltage can be more utilized.

5.2.3 Minimum Available Discontinuous Linear Modulation Index

The duration of non-zero vectors can also become critical in applications with low modulation indices when the discontinuous modulation is used. From table 5.2 it can be seen that the duration of $\vec{V}_{1(pnn)}$ can become critical for the ZVS commutation of phase current I_b. As the reference voltage vector $\vec{V}_{ref}$ approaches $\vec{V}_{2(ppn)}$ in sector I, the duty cycle of $\vec{V}_{1(pnn)}$ becomes shorter while the duty cycle of $\vec{V}_{2(ppn)}$ becomes longer. The duty cycle of nonzero vectors also become shorter with the reduction of the modulation index. As a result the duty cycle of $\vec{V}_{1(pnn)}$ can become critical in low modulation indices near the angle at which the switching sequence changes. Note, that when the phase current I_b becomes positive, i. e. $I_a > I_b > 0 > I_c$, the switching sequence changes to $S_2 \rightarrow S_1 \rightarrow S_0 - S_0 \rightarrow S_1 \rightarrow S_2$ for the remaining of sector I (see also table 4.4 in chapter 4). After changing the switching sequence $\vec{V}_{1(pnn)}$ is no longer critical, even though its duration becomes shorter because the zero vector coming after

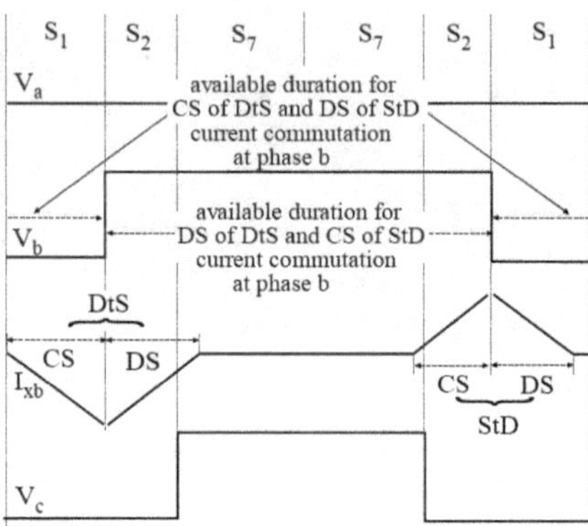

Fig. 5.10: Phase voltages V_a, V_b and V_c and the auxiliary current I_{xb} at phase b *during one switching cycle for discontinuous modulation when I_b is small.*

it provides enough time for the ZVS commutation in phase b.

Another important point, which is worthy to mention is this fact that in the vicinity of the boundary of the current sectors a and b (see Fig. 5.3), the phase current I_b becomes so small that the auxiliary branch of phase b has to be activated to accelerate the commutation of I_b in the transition $S_2 \rightarrow S_1$ of the switching sequence $S_1 \rightarrow S_2 \rightarrow S_7 - S_7 \rightarrow S_2 \rightarrow S_1$ used in the second half of the current sector a. Fig. 5.10 shows this situation. In this case, the auxiliary current of branch b starts to decrease when the voltage polarity at phase b changes, i. e. after the circuit is transferred to switching state S_1. Therefore extra time is necessary for this discharging state, which need to be finished before the charging state of ZVS commutation of I_b for the next transition $S_1 \rightarrow S_2$ begins. This duration which in fact belongs to the transition $S_2 \rightarrow S_1$ should also be considered. But on the other hand, the charging state of ZVS commutation of I_b in the transition $S_1 \rightarrow S_2$ needs less time as the magnitude of I_b becomes smaller. Thus, for the calculation of the necessary duration of the charging state $\Delta t'_{ch}$ related to the transition $S_1 \rightarrow S_2$ $|I_b| = 0.5\,\hat{I}$ is assumed, so the necessary time for the discharging state related to the transition $S_2 \rightarrow S_1$ is also considered with a good approximation. Thus, the condition which limits the minimum duration of the non-zero voltage vectors can be given by:

$$2\,T_1\Big|_{\phi=\theta} > \Delta t'_{ch}, \quad \begin{cases} \theta = \dfrac{\pi}{6} - \dfrac{2\,\pi}{f_S} + \varphi & \text{for} \quad 0 < \varphi < \dfrac{\pi}{6} \\ \theta = \dfrac{\pi}{3} - \dfrac{2\,\pi}{f_S} & \text{for} \quad \dfrac{\pi}{6} < \varphi < \dfrac{\pi}{3} \end{cases} \tag{5.21}$$

where θ is the angle at which the switching sequence changes. From a certain minimum discontinuous modulation index to the lower values, the sum of durations of two vectors $\vec{V}_{1(pnn)}$ placed side by side cannot provide enough time for $\Delta t'_{ch}$. Thereafter, the modulation should be switched to the continuous modulation. In the continuous modulation $\vec{V}_{0(nnn)}$ is also switched between them and provides more time for the mentioned charging state. Introducing T_1 from (4.3) in (5.21) results in:

$$\begin{cases} m_{dm,min} = \dfrac{f_S\,\Delta t'_{ch}}{\sqrt{3}\,\sin\left(\dfrac{\pi}{6} + \dfrac{2\,\pi}{f_S} - \varphi\right)} & \text{for} \quad 0 < \varphi < \dfrac{\pi}{6}, \\ m_{dm,min} = \dfrac{f_S\,\Delta t'_{ch}}{\sqrt{3}\,\sin\left(\dfrac{2\,\pi}{f_S}\right)} & \text{for} \quad \dfrac{\pi}{6} < \varphi < \dfrac{\pi}{3}. \end{cases} \tag{5.22}$$

For $m < m_{dm,min}$, the modulation is changed to the continuous modulation in some parts of every voltage sector. For $\frac{\pi}{6} < \varphi < \frac{\pi}{3}$ these parts are located around the non-zero voltage vectors and for $0 < \varphi < \frac{\pi}{6}$ they are around the angle of the switching sequence changing θ.

Introducing $\Delta t'_{ch}$ from (3.46) and (3.77) for the ARC-HW-ARCP[2] and ARC-QW-ARCP circuits in (5.23), while assuming $|I_b| = 0.5\,\hat{I}$ and $\varphi = 0$ yields:

$$\begin{cases} m_{dm,HW,min} = \dfrac{\sqrt{3}\,f_S\,Z_r\,\hat{I}}{\omega_r\,V_{dc}}, \\ m_{dm,QW,min} = \dfrac{\sqrt{3}}{3}\,\dfrac{f_S\,Z_r\,\hat{I}}{\omega_r\,V_{dc}}, \end{cases} \tag{5.23}$$

[2] $I_{boost} = 0.5\,|I_b|$ is assumed.

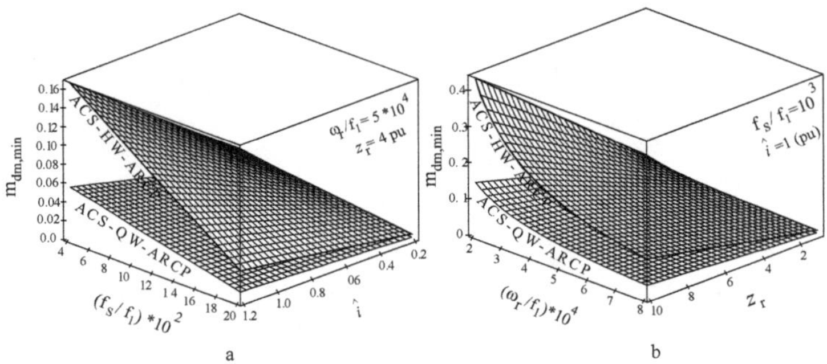

Fig. 5.11: $m_{dm,min}$ for the both commutated pole concepts a) versus the auxiliary circuit characteristics (z_r and ω_r), b) versus f_S and $\hat{i}$.

$m_{dm,HW,min}$ and $m_{dm,QW,min}$ are the respective minimum discontinuous modulation indices. Normalizing the above expressions to the same base values as used before in this chapter results in:

$$\begin{cases} m_{dm,HW,min} & = \dfrac{\sqrt{3}}{2}\dfrac{f_S}{f_1}\dfrac{z_r\,\hat{i}}{(\omega_r/f_1)}, \\ m_{dm,QW,min} & = \dfrac{\sqrt{3}}{6}\dfrac{f_S}{f_1}\dfrac{z_r\,\hat{i}}{(\omega_r/f_1)}. \end{cases} \tag{5.24}$$

Fig. 5.11 shows the dependence of both $m_{dm,HW,min}$ and $m_{dm,QW,min}$ on the auxiliary resonant circuit, z_r and ω_r, sampling frequency ratio, f_S/f_1, and the normalized amplitude of output current, $\hat{i}$.

The angle α_c, within which only the continuous modulation can be used, is given for $\varphi = 0$ by: (see also Fig. 5.6)

$$\alpha_c = 2\arcsin\left(\frac{m_{dm,min}}{2m}\right) - \frac{\pi}{3} \qquad \text{for } m < m_{dm,min} \text{ and } \varphi = 0. \tag{5.25}$$

For $m < m_{dm,min}/\sqrt{3} = 0.577\,m_{dm,min}$ only the continuous modulation can be used. It should also be mentioned that there exist no limitation on $m_{cm,min}$ because the zero voltage vectors being switched at the beginning and at the end of the sampling cycle provide enough time for ZVS commutations.

5.2.4 Available Linear Modulation Area

As discussed in the last sections, the existing limitations on the maximum and minimum modulation indices depend on the sampling frequency and the amplitude of phase current as well as the auxiliary circuit characteristics and topology. Because of these limitations the whole voltage hexagon area cannot be fully utilized. These limitations have been described by expressions (5.13), (5.17) and (5.23). A maximum available linear modulation index lower than that of hard switching converters, i.e. $m_{max} = 1.15$, means a reduction in the linearity range of modulation. Fig. 5.12 illustrates for the HW- and QW-circuits with given parameters the available linear modulation area of the voltage hexagon using the continuous and discontinuous modulation techniques when $\varphi = 0$. The maximum and minimum values of the modulation indices for each case is obtained from expressions mentioned above.

As indicated from the figure, the discontinuous modulation technique is advantageous in operations with high modulation indices because it offers a higher maximum linear modulation index. The resulting reduction in linearity range of SVM using this kind of modulation is considerably less than that of the continuous modulation. In contrary the continuous modulation would be preferable at low modulation index operations, because there is no limitation on the minimum modulation index of this technique. The ARC-QW-ARCP circuit also offers a wider linearity range for the both modulation techniques. This is because of full DC link voltage applied across the inductors of its auxiliary circuit during the charging and discharging states.

5.3 Summary of Chapter

The duration of three stages of Zero Voltage Transition (ZVT), namely charging, resonant and discharging states, as well as duration of current self-commutation (see appendix D) affect the DC bus utilization of soft commutated converters and reduce the linearity range of Space Vector PWM (SVPWM). This chapter shows how the durations of transient stage of ZVT and of current self-commutation reduce the DC bus utilization. This reduction becomes more considerable in Half-Wave (HW) concept (see appendix D) because the transient state of ZVT in this concept takes more time compared to Quarter-Wave (QW) one under the same resonant circuit characteristics. The DC bus utilization decreases also with the increase of the switching frequency. An effective method, which can be simply implemented without any additional hardware, is suggested to compensate this reduction.

The necessary duration of discharging stage of ZVT restricts the minimum duration of zero voltage vectors leading to reduction of linearity range of modulation. The expressions describing the maximum available linear modulation indices for both Half- and Quarter-Wave concepts as well as for continuous and discontinuous modulation techniques are derived. The maximum available linear modulation index becomes smaller as the switching frequency or load current

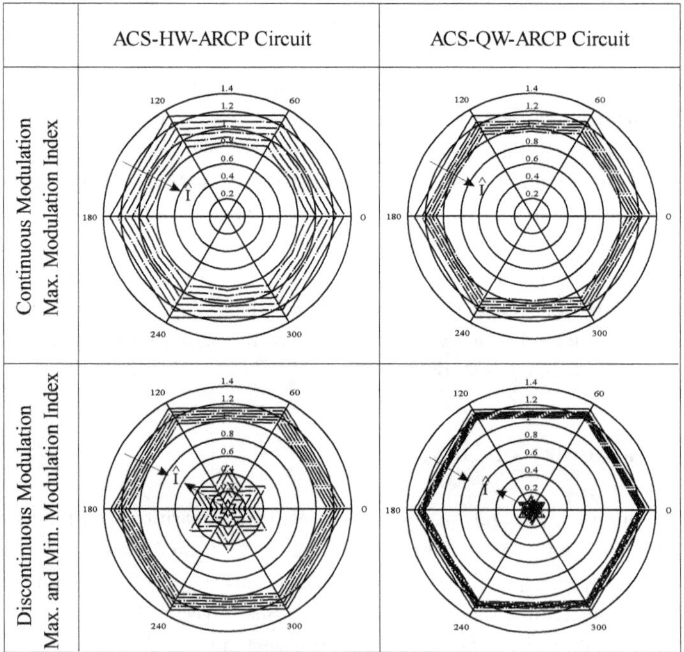

Fig. 5.12: Available linear modulation area limited by the available maximum and minimum modulation index (f_S=60KHz, φ=0, Z_r=24.7Ω, ω_r=2.5MHz and $\hat{I}$=0...100A)

increases. In high power high frequency applications it can become even smaller than one.

The duration of charging stage of ZVT limits the minimum duration of non-zero voltage vectors causing the minimum available linear modulation index to be also limited for discontinuous modulation techniques.

Due to the mentioned restrictions soft commutated PWM converters are not able to make the maximum linear modulation index of conventional hard commutated converters, $m_{max} = 1.15$, available. The results show that the reduction of linearity range of modulation is more considerable with the Half-Wave topologies. The same is true when using the conventional continuous space vector PWM.

6. A DSP-Based Controller for ASSC PWM Pole Converters

The issues of the most advantageous ASSC PWM circuit and the optimized modulation method have been discussed in the previous chapters. The last issue which should be discussed in this thesis is to find a proper current control method. Selecting the commutated pole concept of the ASSC PWM converters as the best choice and employing the ML-SVPWM switching technique to achieve minimum losses in the converter, it is now to decide which current control techniques facilitate the previous choices. By selecting the ML-SVPWM technique the switches of the circuits have to be then switched according to a certain switching pattern. It is not possible under this circumstance to use the direct current control techniques, for example those presented in [84]-[90]. In the direct techniques the switching state at each instant is directly determined by the current regulator in such a way that transient and/or steady state behaviour of the converter and current controller is optimized. Hence, the degrees of freedom of PWM are not available to be utilized by a certain modulation technique for optimizing other criteria such as converter losses as discussed in last chapter. The only current controller which jointly operates with the SVPWM and allows the degrees of freedom of modulation to be available for optimizing the converter losses is the space vector based rectifier regulator for AC/DC/AC converters introduced in [33]. This technique is known as "Dead-Beat Current Control Technique" and commonly used for control of the input current of the source side converter of AC/DC/AC converter systems as shown in Fig. 6.1.

As can be seen from Fig. 6.1 the AC source voltages have to be introduced into the Dead-Beat current controller. Utilizing the source voltage informations, the controller can drive the line current to the desired value over one sampling cycle in steady state operations, so the converter is able to exactly regulate the power transferred and correct the power factor at the same time as well. The

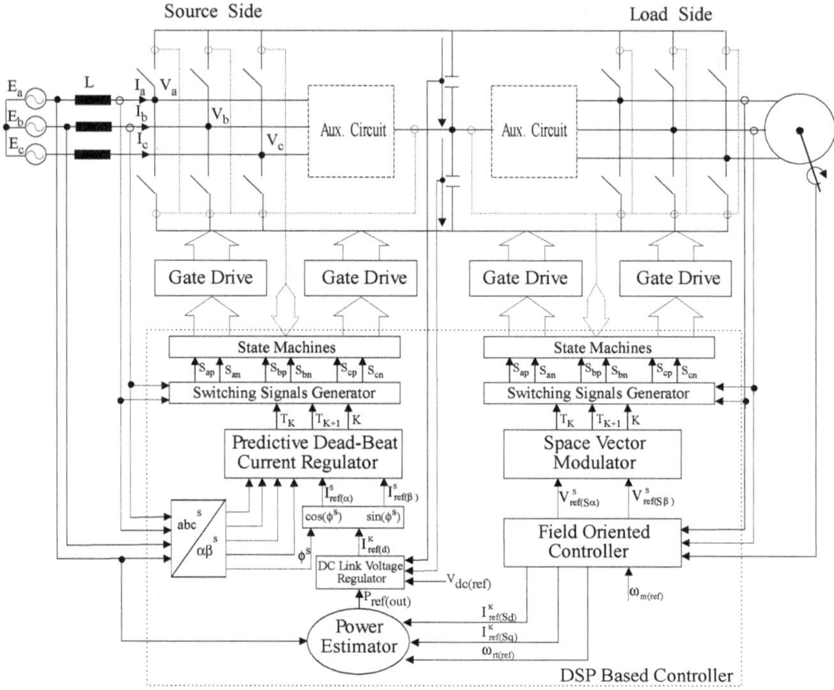

Fig. 6.1: Three-Phase AC/DC/AC Converter System.

principle of this controller at steady state and transient operations, while the source and load side units of the DSP-based controller of Fig. 6.1 are explained, will be discussed in this Chapter.

6.1 Source Side Part of the DSP Controller

The three phase source voltages assumed to be balanced, sinusoidal with the frequency ω_1 can be described in a three phase system as follows:

$$E_a(t) \quad = \quad \hat{E}\cos(\omega_1 t), \tag{6.1}$$

$$E_b(t) = \hat{E}\cos(\omega_1 t - \frac{2\pi}{3}), \tag{6.2}$$

$$E_c(t) = \hat{E}\cos(\omega_1 t + \frac{2\pi}{3}). \tag{6.3}$$

These voltages are given in the stationary $\alpha\beta$ reference frame by:

$$\vec{E}^S = E^S_\alpha + j\,E^S_\beta = E_a + j\left(\frac{E_a + 2\,E_b}{\sqrt{3}}\right) = \hat{E}\,cos(\omega_1 t) + j\,\hat{E}\,\sin(\omega_1 t). \tag{6.4}$$

$\vec{E}^S$ is described in the dq reference frame rotating with the frequency ω_1 and aligned with E_a by:

$$\vec{E}^K = E^S_\alpha\,cos(\omega_1 t) + E^S_\beta\,\sin(\omega_1 t) = \hat{E}. \tag{6.5}$$

The source currents are also described in these reference frames by:

$$\vec{I}^S = I^S_\alpha + j\,I^S_\beta = I_a + j\left(\frac{I_a + 2\,I_b}{\sqrt{3}}\right), \tag{6.6}$$

$$\begin{aligned} \vec{I}^K &= I^K_d + j\,I^K_q = \Big(I^S_\alpha\,\cos(\omega_1 t) + I^S_\beta\,\sin(\omega_1 t)\Big) \\ &\quad + j\Big(-I^S_\alpha\,\sin(\omega_1 t) + I^S_\beta\,\cos(\omega_1 t)\Big) = I^K_d + j\,I^K_q. \end{aligned} \tag{6.7}$$

In order to use the source side converter as a power factor corrector and to obtain unity power factor, the q-component of reference current vector, $\vec{I}^K_{ref(q)}$, oriented in phase with E^K is set to zero:

$$\vec{I}^K_{ref(q)} = 0. \tag{6.8}$$

Consequently, the d-component $I^K_{ref(d)}$ corresponds to the amplitude of the input current and should be so regulated that the transferred power covers both the converter losses and the power demanded by the load.

The power demanded by the load can be estimated on-line from the load currents and other parameters (machine parameters) in a power estimator unit (see Fig. 6.1) as follows:

$$P_{ref(out)} = P_{Load} + P_{Losses} = \frac{3}{2}\left\{\left[(r_s + r_r\left(\frac{L_m}{L_r}\right)^2\right] I_{Sd}^{K^2} + r_s\, I_{Sq}^{K^2} + \frac{L_m^2}{L_r}\, I_{Sd}^K\, I_{Sq}^K \omega_{rt(ref)}^K\right\} + 6\, V_{drop}\, |I_a|. \quad (6.9)$$

The first part estimates the power delivered to the load and the second part takes roughly the losses of the semiconductors into account. From the power balance concept, (6.5) and (6.8) it results:

$$P_{in} = \frac{3}{2}\,(E_d^K\, I_d^K + E_q^K\, I_q^K) = \frac{3}{2}\,\hat{E}\, I_d^K = P_{out}, \quad (6.10)$$

and from there:

$$I_d^K = \frac{2}{3}\frac{P_{out}}{\hat{E}}. \quad (6.11)$$

The desired value of $I_{ref(d)}^K$ is given by the DC link voltage regulator, which is a simple PI-regulator and regulates the DC bus voltage, as shown in Fig. 6.1. It follows:

$$I_{ref(d)}^K = \frac{2}{3}\frac{P_{ref(out)}}{\hat{E}} + \frac{1}{R_s}\left[K_p(V_{dc(ref)} - V_{dc}) + K_i \int (V_{dc(ref)} - V_{dc})dt\right], \quad (6.12)$$

R_s is a trade off parameter for weighing the steady-state or transient performances. If R_s is chosen small, the DC link voltage regulation is dominant, and the steady-state DC link voltage ripple is low. If R_s is selected large, the transient performance is improved.

The current reference, $I_{ref(d)}^K$, is then transformed back to the $\alpha\beta$ stationary reference frame and given by:

$$\vec{I}_{ref}^S(t) = I_{ref(\alpha)}^S + I_{ref(\beta)}^S = I_{ref(d)}^K\, cos(\omega_1\, t) + j\, I_{ref(d)}^K\, \sin(\omega_1\, t). \quad (6.13)$$

The reference current vector, $\vec{I}_{ref}^S$, is used by the dead-beat current regulator described in the next section for calculating the necessary voltage at the AC side of the converter.

6.1.1 Predictive Dead-Beat Current Regulator

6.1.1.1 Steady State Operation

The current regulator unit calculates the reference voltage vector $\vec{V}^S_{ref}$ which should be generated at the AC side of the converter to drive the line current vector, $\vec{I}^S$, to the current reference vector, $\vec{I}^S_{ref}$, so that $\vec{I}^S$ reaches $\vec{I}^S_{ref}$ within one sampling cycle T_S. The required voltage reference vector, $\vec{V}_{ref}$, is then realized at the AC side of the converter using the ML-SVPWM technique described in chapter 4.

For the calculation of the reference voltage, the variation of the line current within the sampling cycle is assumed linear. For the nth sampling cycle, the variation of the line current can be then described in the stationary reference frame as follows:

$$\begin{aligned} \Delta\vec{I}^S\Big(nT_S\Big) &= \vec{I}^S\Big((n+1)T_S\Big) - \vec{I}^S\left(nT_S\right) \\ &= \frac{T_S}{L}\Big(\vec{E}^S\left(nT_S\right) - \vec{V}^S_{ref}\left(nT_S\right)\Big). \end{aligned} \tag{6.14}$$

The regulation principle demands that at the end of the sampling cycle the line current vector, $\vec{I}^S$, reaches the desired reference vector, $\vec{I}^S_{ref}$, given by the DC voltage regulator at the beginning of the cycle. This is:

$$\vec{I}^S\Big((n+1)T_S\Big) = \vec{I}^S_{ref}(nT_S). \tag{6.15}$$

Introducing (6.15) into (6.14) results in:

$$\vec{I}^S_{ref}(nT_S) - \vec{I}^S(nT_S) = \frac{T_S}{L}\left(\vec{E}^S(nT_S) - \vec{V}^S_{ref}(nT_S)\right). \tag{6.16}$$

$\vec{V}^S_{ref}$ is then calculated as follows:

$$\vec{V}^S_{ref}(nT_S) = \vec{E}^S(nT_S) - \frac{L}{T_S}\left(\vec{I}^S_{ref}(nT_S) - \vec{I}^S(nT_S)\right). \tag{6.17}$$

If the source voltage, $\vec{E}^S$, is assumed constant within the sampling cycle, the generation of $\vec{V}^S_{ref}$ at the AC side of the converter guarantees that $\vec{I}^S(nT_S)$ reaches

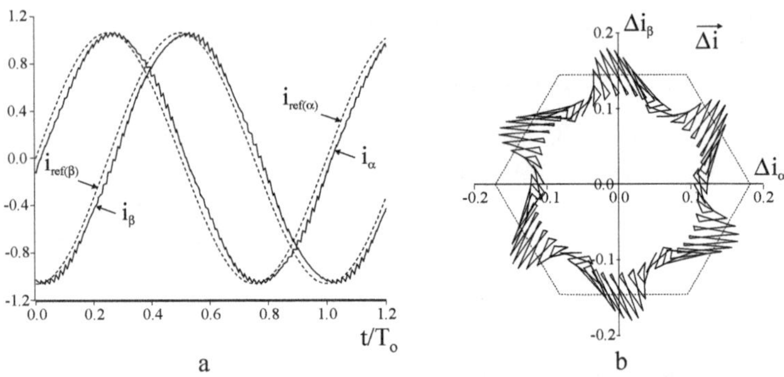

Fig. 6.2: a) Normalized α- and β-components of the AC side current and reference current and b) locus curve of normalized current-error vector trajectory in αβ-plane for the dead-beat regulator.

the reference current, $\vec{I}^S_{ref}(nT_S)$, at the end of the sampling cycle.

Once the required voltage vector, $\vec{V}^S_{ref}(nT_S)$, for the nth sampling cycle is calculated, it can be generated using ML-SVPWM technique. The information about the number of the voltage sector K as well as duty cycles T_K and T_{K+1} are used in the Switching Signal Generator unit to generate the initial switch on/off signals according to the switching sequence of the ML-SVPWM technique (see Fig. 6.1).

Since with this regulation method the line current reaches its reference value with a delay as much as one sampling cycle, it is called the "Dead-Beat Current Regulation". The dead-time can be clearly seen in Fig. 6.2(a) from the existing phase difference between the waveforms of line current components and their respective reference waveforms. This can also be recognized from the empty circle built in the middle of the locus curve of current-error vector trajectory in Fig. 6.2(b) corresponding to a dead-time function. This is also worth mentioning that the magnitude of the current-error vector in Fig. 6.2(b) becomes maximum (locally) at the middle of every voltage sector, because the duty cycles T_K and T_{K+1} become equal at these positions.

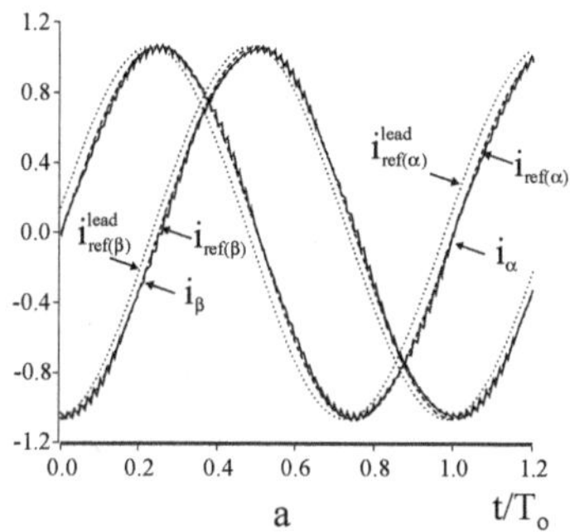

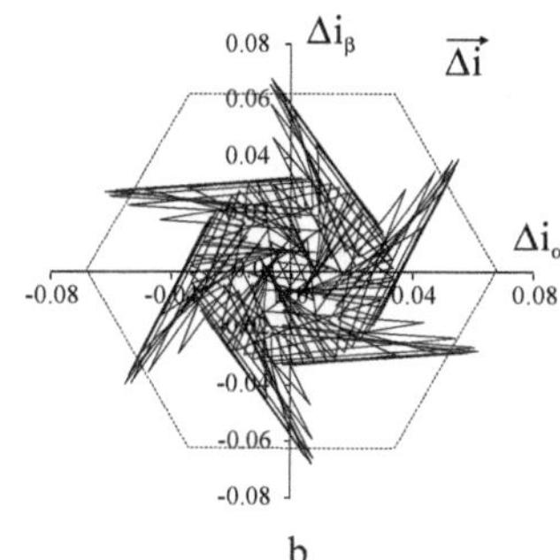

Fig. 6.3: a) Normalized α- and β-components of the phase current and reference current vectors and b) locus curve of normalized current-error vector trajectory in $\alpha\beta$-plane for a predictive dead-beat regulator.

The results shown in Fig. 6.2 reveal that the magnitude of the current-error will never become zero because of the aforementioned delay. In order to avoid this delay, the current reference value of the next sampling cycle is predicted and used instead of the actual one. In this way, the voltage reference vector resulting from (6.17) drives the line current exactly to the current reference predicted for the next sampling cycle. This strategy is known as "Predictive Dead-Beat Current Regulation". The results of the predictive mode of this regulator are shown in Fig. 6.3. If the waveform of the reference current is known such as the example of this section, it can be simply shifted leading by one sampling cycle as shown in Fig. 6.3(a) (marked by $\vec{I}_{ref}^{lead}$). From Fig. 6.3(b) is visible that the steady state current-error becomes zero in the predictive mode of operation. Another interesting point in this figure is the maximum magnitude of current-error vector, $\Delta\vec{i}$, which is significantly reduced in this mode.

6.1.1.2 Transient Operation

Transient operation can be defined as an operation that the solution of equation (6.17) results in a voltage vector located outside of the space vector voltage hexagon and therefore cannot be synthesized by $\vec{V}_K$ and $\vec{V}_{K+1}$ over the sampling cycle T_S. Under this condition, dead-beat control of the line current is no longer possible. In this case the best transient performance which can be achieved is to

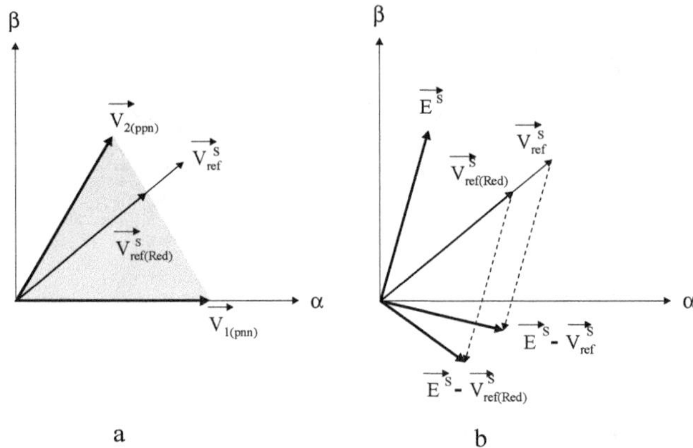

Fig. 6.4: a) Proportional reduction of voltage reference and b) Incorrect direction of $\Delta\vec{I}$ *caused by* $\vec{V}^S_{ref(Red)}$.

switch the converter in such a way that the current vector $\vec{I}^S$ is driven in the *direction* of the reference vector $\vec{I}^S_{ref}$. For this purpose, the angle of the actual current-error vector, $\Delta\vec{I}(nT_S) = \vec{I}^S\Big((n+1)T_S\Big) - \vec{I}^S(nT_S)$, should be kept equal to the angle of reference current-error vector, $\Delta\vec{I}_{ref}(nT_S) = \vec{I}^S_{ref}\Big((n+1)T_S\Big) - \vec{I}^S(nT_S)$. Generating a proportionally reduced voltage vector, as shown in Fig. 6.4(a), does not drive the line current vector in the correct direction. It can be easily understood from equations (6.18) and (6.19) and from Fig. 6.4(b) as well.

$$\begin{aligned} \Delta\vec{I}_{Red}(nT_S) &= \vec{I}^S\Big((n+1)T_S\Big) - \vec{I}^S(nT_S) \\ &= \frac{T_S}{L}\left(\vec{E}^S(nT_S) - \vec{V}^S_{ref(Red)}(nT_S)\right), \end{aligned} \tag{6.18}$$

$$\begin{aligned} \Delta\vec{I}^S_{ref}(nT_S) &= \vec{I}^S_{ref}\Big((n+1)T_S\Big) - \vec{I}^S(nT_S) \\ &= \frac{T_S}{L}\left(\vec{E}^S(nT_S) - \vec{V}^S_{ref}(nT_S)\right). \end{aligned} \tag{6.19}$$

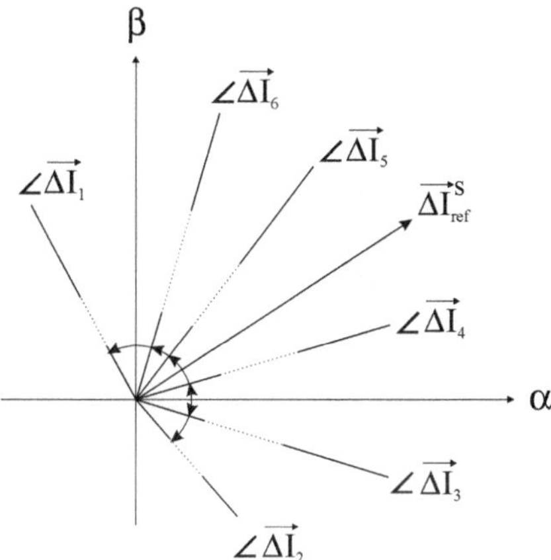

Fig. 6.5: Selection of the two $\Delta\vec{I}_K$ adjacent to $\Delta\vec{I}^S_{ref}$ for transient operations.

To find the appropriate voltage vectors, $\vec{V}_K$ and $\vec{V}_{K+1}$, which drive $\Delta\vec{I}^S$ in the direction of $\Delta\vec{I}^S_{ref}$, the angle of the current-error vectors, $\angle\Delta\vec{I}_K(nT_S)$, related to all six non-zero voltage vectors should be calculated as follows:

$$\angle\Delta\vec{I}_K(nT_S) = \angle\Big(\vec{E}^S(nT_S) - \vec{V}_K\Big) \quad K = 1, \cdots, 6. \tag{6.20}$$

The two current-error vector angles $\angle\Delta\vec{I}_K$ and $\angle\Delta\vec{I}_{K+1}$ being adjacent of $\Delta\vec{I}^S_{ref}$ correspond to the voltage vectors $\vec{V}_K$ and $\vec{V}_{K+1}$ driving the line current in the desired direction. Fig. 6.5 shows an example in which the voltage vectors $\vec{V}_{4(npp)}$ and $\vec{V}_{5(nnp)}$ are proposed.

Since $\Delta\vec{I}_K$ and $\Delta\vec{I}_{K+1}$ found by this method are adjacent of the desired direction, switching of the respective voltage vectors $\vec{V}_K$ and $\vec{V}_{K+1}$, each one for a certain duration, drives $\vec{I}^S$ in the desired direction. The duty cycles of the voltage vectors T_K and T_{K+1} can then be calculated from the following equation:

$$\Delta\vec{I}^S_{ref}(nT_S) = \frac{T_K}{L}\Big(\vec{E}^S(nT_S) - \vec{V}_K\Big) + \frac{T_{K+1}}{L}\Big(\vec{E}^S(nT_S) - \vec{V}_{K+1}\Big). \tag{6.21}$$

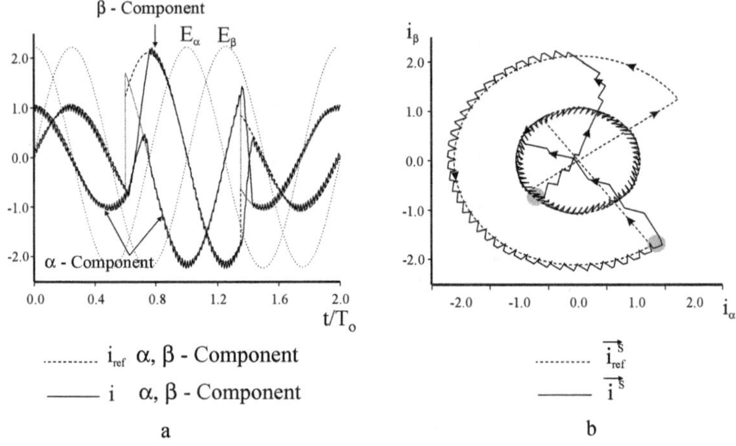

Fig. 6.6: Behaviour of Dead-Beat Regulation at the transient state.

This is a polar equation resulting in two scalar equations. Since the sum of T_K and T_{K+1} is now larger than the sampling cycle T_S, they are proportionally reduced according to (6.22) and (6.23) without changing the angle of $\Delta\vec{I}^{S}$[1] .

$$T'_K = T_K \cdot \frac{T_S}{T_K + T_{K+1}}, \tag{6.22}$$

$$T'_{K+1} = T_{K+1} \cdot \frac{T_S}{T_K + T_{K+1}}. \tag{6.23}$$

Figure 6.6 illustrates the behaviour of the current regulator at the transient state. It is important to note that the transition duration strongly depends on the position of $\vec{E}^S$. In Fig. 6.6(b) it can be clearly seen that the first transition takes twice as much time as the second one. The reason for it can be understood from the Fig. 6.6(a). In the first transition both current components should change their direction from negative to positive (see the current directions in Fig. 6.1) while both components of source voltage E_α and E_β are large and negative. Hence, the converter has to generate a larger negative voltage in α and β directions.

[1] As if both sides of equation (6.21) are divided by a constant.

In this situation, only the voltage sector IV can be selected and the modulator has to switch between $\vec{V}_{4(npp)}$ and $\vec{V}_{5(nnp)}$. Switching of only $\vec{V}_{5(nnp)}$ generates an α-component voltage as much as $(-V_{dc}/3)$ which is not sufficient to reverse the direction of I_α at this time. Also switching of only $\vec{V}_{4(npp)}$ does not drive the current vector in the correct direction because it generates no β-component voltage at the AC side. The situation is even worse for the β-component of the line current because it has to jump to its positive maximum value, while E_β is increasing to its negative peak. The regulator should therefore wait until $E_\alpha > (-V_{dc}/3)$, then the voltage vector $\vec{V}_{5(nnp)}$ can be switched for the most time. This is also visible in Fig. 6.6(b). These conditions are different altogether for the second transition. In this transition I_α should become negative, while E_α is near its negative peak. It can simply be accomplished by generating a small positive voltage in the α-direction. Also I_β should become positive, while E_β is positive and large but it is also decreasing towards zero. Thus, some negative voltage is necessary in β-direction. By selecting the voltage sector VI and switching between $\vec{V}_{1(pnn)}$ and $\vec{V}_{6(pnp)}$ the regulator can optimize the duration of transition more easily.

It is also worth noting, that at transients states the prediction of current reference causes the line current to follow its steady state reference value for one further sampling cycle after the current reference step occurs. This is shown in Fig. 6.6(b) with gray circles.

The switching signal generator and state machine units are the same for the source and load sides. They will be described in the following section.

6.2 Load Side Part of the DSP Controller

The load side current controller is the well-known Field Oriented Controller with decoupling the rotor flux. It is not the object of this thesis to discuss this kind of current controller. For the detailed description of this controller see [9] and [10].

The current controller provides the necessary informations about the required voltage at the load terminals. A modulator uses the informations to determine the number of voltage sector and the duty cycles of the voltage vectors which are subsequently used by the switching signal generator unit to create the initial

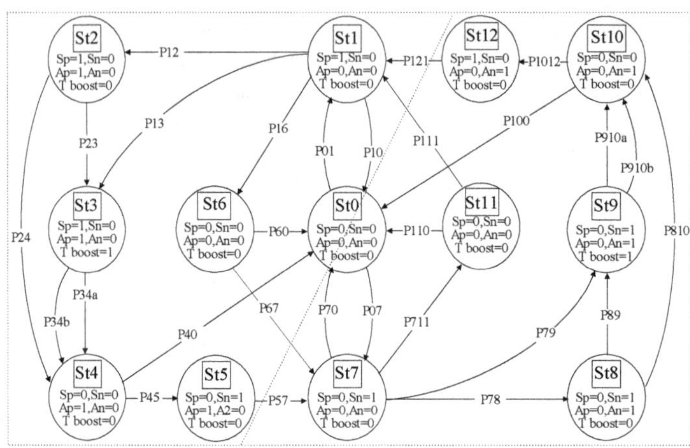

Fig. 6.7: State machine for ACS-HW-ARCP converter system.

turn-on/off signals for the main switches according to the switching sequence of the ML-SVPWM technique. The turn-on/off signals related to each phase are then used in the respective state machine unit producing the final turn-on/off signals for the main and auxiliary switches.

The state machine unit of each phase recognizes the kind of current commutation at that phase when a switch should be turned on/off. It also ensures the complete ZVS commutation by observing the currents of the upper and bottom diodes of the respective phase. The state machine of the ACS-HW-ARCP converter is shown in Fig. 6.7. Each phase has its own state machine operating independent from the other phases. The order of state changing is the same for all phases and for both the source and load side converters. The state changes, only when the condition function of the path between the origin and destination states becomes true. The paths and their condition function as well as the signals being observed in the origin state for detecting a possible change between two states are given in the table 6.1.

Path Conditions for HW-concept											
Path	*Controller Signals*				*Measured Signals*						*Bowl Function*
	Sp	*Sn*	*Tout*	*Mode*	*Spon*	*Snon*	*Dpon*	*Dnon*	*Zc*	*St*	*Bf*
P01	**1**									**1**	$St\&Sp$
P07		**1**								**1**	$St\&Sn$
P12	**0**				**0**					**1**	$St\&\overline{Sp}\&\overline{Spon}$
P13	**0**			**1**	**1**					**1**	$St\&\overline{Sp}\&$ $Spon\&Mode$
P16	**0**			**0**	**1**					**1**	$St\&\overline{Sp}\&$ $Spon\&\overline{Mode}$
P23					**1**					**1**	$St\&Spon$
P24										**0**	$\overline{St}$
P34a			**1**							**1**	$St\&Tout$
P34b										**0**	$\overline{St}$
P40									**1**	**0**	$\overline{St}\&Zc$
P45								**1**		**1**	$St\&Dnon$
P57									**1**	**1**	$St\&Zc$
P60										**0**	$\overline{St}$
P67								**1**		**1**	$St\&Dnon$
P70										**0**	$\overline{St}$
P78		**0**				**0**				**1**	$St\&\overline{Sn}\&\overline{Snon}$
P79		**0**		**1**		**1**				**1**	$St\&\overline{Sn}\&$ $Snon\&Mode$
P711		**0**		**0**		**1**				**1**	$St\&\overline{Sn}\&$ $Snon\&\overline{Mode}$
P89						**1**				**1**	$St\&Sn$
P810										**0**	$\overline{St}$
P910a			**1**							**1**	$St\&Tout$
P910b										**0**	$\overline{St}$
P100									**1**	**0**	$\overline{St}\&Zc$
P1012							**1**			**1**	$St\&Dpon$
P110										**0**	$\overline{St}$
P111							**1**			**1**	$St\&Dpon$
P121									**1**	**1**	$St\&Zc$

Table 6.1: Path conditions between the states of state machine for ACS-HW-ARCP converter system.

The signals shown inside the state circles in Fig. 6.7 are the final output signals for the gate drive circuits. The states 1 to 6 corresponding to the current commutation from the upper branch to bottom branch of each phase are described below. The states 7 to 12 have the same functions for the current commutation from the bottom branch to upper branch of the phase.

The symbols used in the state changing flowchart and path condition table are introduced as follows:

- St: start signal 1: start and 0: stop.
- Sp: Initial gate signal of upper switch.
- Sn: Initial gate signal of bottom switch.
- Spon: Conduction of upper switch.
- Snon: Conduction of bottom switch.
- Dpon: Conduction of upper diode.
- Dnon: Conduction of bottom diode.
- Tout: End of boost charging duration.
- T-boost: Boost charging signal 1: begin and 0: end.
- Zc: Zero-crossing signal for auxiliary inductor current.
- Mode: Auxiliary circuit extension mode for switch-to-diode current commutations with small current magnitude.

When the circuit is switched on and, for example, the input initial signal *Sp* becomes one, the path "P01" becomes true and changes the state from "ST0" to "ST1", thereby the output signal "Sp" also becomes active and the upper switch is turned on. When this signal changes again to zero at input and the antiparallel diode of the upper switch is conducting (signal "*Spon*" is zero), the path "P12" becomes true, thereby the state changes to "ST2", the respective auxiliary switch (signal "AP") is turned on and a ZVS diode-to-switch current commutation process begins. The state "St2" corresponds to the charging state of the auxiliary branch. At the end of this state the auxiliary current exceeds the phase current causing the upper switch to start conducting. This changes the signal "*Spon*" to one and makes the path "P23" true resulting in changing the state to "St3" and beginning the boost charging state. The signal "T-boost", which also becomes active in this state, actuates a count-down counter in the switching signal generator. This count-down counter keeps the necessary boost charging time precalculated in the signal generator from the informations about the magnitude of phase current. At the end of this time the signal generator activates the signal "T_{out}" which makes the path "P34a" true and changes the state to "St4". The signal "T-boost" becomes again zero and resets the count-down counter to the waiting position. In this state the output signal "Sp" becomes zero, the upper switch is turned off and the resonant state begins. When the phase voltage reaches the negative DC rail and the bottom diode starts conducting, the signal "D_{non}" becomes active and changes the state to "St5" through path "P45", the bottom switch is turned on under zero voltage and the discharging state begins. When this state ends and the zero-crossing of the auxiliary inductor current is detected through signal "*Zc*", state "St7" is switched, thereby the auxiliary switch is turned off and the current commutation from the upper diode to the bottom switch ends.

In case upper switch is conducting (signal "*Spon*" is one) when input signal "*Sp*" changes from one to zero in state "St1" and the phase current is large enough (signal "*Mode*" is zero), path "P16" becomes true and the state "St6" follows. "Sp" at the output becomes zero and the upper switch is turned off. A self-commutation occurs which leads to changing of the polarity of the phase voltage. When the bottom diode starts conducting, the state changes to "St7" and the bottom switch is turned on. If the magnitude of phase current is not large enough for a self-commutation, the signal generator actuates the signal "*Mode*" triggering the auxiliary branch to amplify the commutating current. In this case the path "P13" becomes true and the state changes from "St1" to "St3". A ZVT commutation process without the charging state, "St2", follows. The further order of events is the same as described for the diode-to-switch current commutation.

Fig. 6.8 illustrates the results of a mixed-mode (digital and analog) simulation of the source side converter with simulator SABER. Fig. 6.8(a) shows the order of states in an extension mode of operation of auxiliary branch related to phase a during a switch-to-diode current commutation. Fig. 6.8(b) shows the order of state changing in a normal mode, while phase current I_a is being commutated

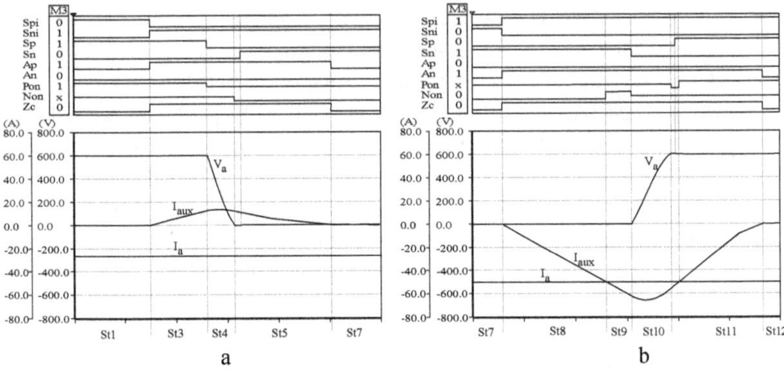

Fig. 6.8: Mixed-mode simulation of the ACS-HW-ARCP converter system: a) extension mode of switch-to-diode current commutation and b) diode-to-switch current commutation.

from the diode of the upper branch to the switch of the bottom branch. In the simulations, the current of each branch is sensed only by one sensor and converted to a three-state digital signal. The three-state signals, "Pon" and "Non" belonging to the upper and bottom branches respectively, become one, zero or undefined (x) in case positive current, negative current or no current is detected which represent conduction of switch, conduction of diode or turn-off state of both devices respectively. The obtained informations are then converted to two two-state digital signals used in the table 6.1. For example, "*Spon*" become one when "Pon" becomes one whereas "*Dpon*" become one when "Pon" becomes zero, otherwise they remain zero. In the simulations, a short delay is also considered for the final gate signals "Sp" and "Sn" in states "St7" and "St12" to ensure the zero voltage conditions at switch turn-on times.

The state machine and its path condition table can be also used in a ACS-QW-ARCP converter system with some simple changes. Fig. 6.9 illustrates the state machine for ACS-QW-ARCP converter system. Its path condition functions are given in table 6.2. Note that in this converter system there is no need for current sensing in the auxiliary branch.

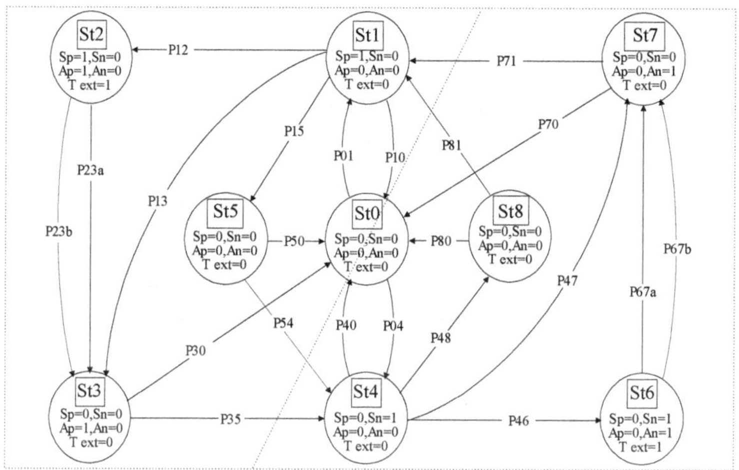

Fig. 6.9: State machine for ACS-QW-ARCP converter system.

6.3 Summary of Chapter

When an optimized modulation technique is used for minimizing the converter switching losses, the switches of converter have to be switched according to the switching pattern determined by the modulation technique. Under this circumstance employing a direct current control method which controls the converter current directly through the switching states is not possible. One of the indirect current control methods which is simple to implement and controls the converter current through the voltage vector generated at the AC-side of converter is the space vector based Dead-Beat Current Control Technique.

Explaining the source and load side units of a DSP-based controller for an AC/DC/AC system, this chapter describes the operation principles of the dead-beat current control technique in both steady and transient states. The state flowchart controlling the switches of main and auxiliary circuits is described for the AC-Side Half/Quarter-Wave Auxiliary Resonant Commutated Pole (ACS-H/QW-ARCP) circuits. Mixed mode simulations are carried out with SABER to test the correctness of state flowcharts.

Path Conditions for QW-concept										
Path	*Controller Signals*				*Measured Signals*					*Bowl Function*
	Sp	*Sn*	*Tout*	*Mode*	*Spon*	*Snon*	*Dpon*	*Dnon*	*St*	*Bf*
P01	**1**								**1**	$St\&Sp$
P07		**1**							**1**	$St\&Sn$
P12	**0**			**1**	**1**				**1**	$St\&\overline{Sp}\&$ $Spon\&Mode$
P13	**0**				**0**				**1**	$St\&\overline{Sp}\&\overline{Spon}$
P15	**0**			**0**	**1**				**1**	$St\&\overline{Sp}\&$ $Spon\&\overline{Mode}$
P23a			**1**						**1**	$St\&Tout$
P23b									**0**	$\overline{St}$
P30									**0**	$\overline{St}$
P34								**1**	**1**	$St\&Dnon$
P40									**0**	$\overline{St}$
P46		**0**		**1**		**1**			**1**	$St\&\overline{Sn}\&$ $Snon\&Mode$
P47		**0**				**0**			**1**	$St\&\overline{Sn}\&\overline{Snon}$
P48		**0**		**0**		**1**			**1**	$St\&\overline{Sn}\&$ $Snon\&\overline{Mode}$
P50									**0**	$\overline{St}$
P54								**1**	**1**	$St\&Dnon$
P67a			**1**						**1**	$St\&Tout$
P67b									**0**	$\overline{St}$
P70									**0**	$\overline{St}$
P71							**1**		**1**	$St\&Dpon$
P80									**0**	$\overline{St}$
P81							**1**		**1**	$St\&Dpon$

Table 6.2: Path conditions between the states of state machine for ACS-QW-ARCP converter system.

7. Summary

The AC Side Soft Commutated (ASSC) PWM Three-Phase Converters are very advantageous for medium and high power applications because their auxiliary circuit located at the AC side of converter have no component in series with the power flow and processes only a small fraction of the total converter power for a short time. For the same reasons an AC-side soft commutated converter benefits from the low losses of its auxiliary circuit. The normal operation of converter also remains functionally identical to its hard commutating counterpart after commutation, since the auxiliary circuit is active only during the current commutation. The auxiliary switches have to withstand either full DC link voltage or even half of it depending on the topology of auxiliary circuit. The maximum current of the auxiliary switches can reach a value twice as large as the amplitude of phase current, but their average current remains small because the auxiliary switches conduct only during the switching transition which is very short compared to the switching cycle.

This work addresses several key issues related to AC-side soft commutated (ASSC) PWM converters. These are:

1. analysis, comparison, and exploration of design equations of ASSC PWM three-phase converter topologies,

2. reexamination of different PWM techniques and development of a modulation algorithm for minimizing the losses of ASSC PWM converters,

3. investigation of influence of voltage transition durations on the DC bus utilization of converters and on the linearity range of modulation techniques,

4. suggestion of a current control method which facilitates the developed modulation algorithm.

Chapter 2: Using SABER simulator program and accurate physics-based dynamic models of IGBT and power diode, the superiority of ZVS technique in reducing the problems relating to hard switching of IGBTs and reverse-recovery effect of power diodes is shown at first. Discussing the AC side soft commutated topologies employing the ZVS technique, this chapter categorizes them in two ways:

1. *Half-Wave (HW) and Quarter-Wave (QW) concepts:* the auxiliary resonant network of Half Wave-topologies provides the ZVS conditions after a half resonance cycle, while the resonant network of Quarter Wave-topologies provides these conditions after a quarter resonance cycle (see also appendix D).

   Comparison: (under the same resonant circuit characteristics) The voltage transition in Quarter Wave-topologies is twice as fast as in Half Wave-topologies. The Quarter Wave-concept also ensures complete ZVS conditions at any load condition, whereas the Half Wave-concept needs a certain amount of energy (boost energy) for providing these conditions depending on the load current. Because ZVT is faster in the Quarter Wave-concept, the reduction of DC bus utilisation and of the linearity range of modulation in this concept is less than that in Half Wave-concept. However, hard turn-off conditions at the auxiliary circuit of Quarter Wave-topologies exclude the implementation of auxiliary switches of this concept with IGBTs. The auxiliary switches of half Wave-topologies are turned on/off under ZCS conditions making use of IGBTs as auxiliary switch possible. This leads to less losses in the auxiliary circuit, thereby operation of converter at a higher switching frequency becomes possible.

2. *Commutated Pole and Non-Pole Commutated concepts:* in commutated pole topologies the soft current commutation occurs independently in every phase. In contrary, in the non-pole commutated converters the soft commutations of phase currents are not independent from each other.

   Comparison: In a commutated pole concept, the converter can be controlled by any modulation technique. This allows the modulation technique to be modified in such a way that some criteria such as the overall losses are optimized. But, in a non-pole commutated concept the converter should be controlled by a certain PWM control algorithm which is developed to meet the requirement of the auxiliary circuit operation and is a suboptimized one concerning converter losses or higher order harmonics. The sampling frequency of reference voltage in a commutated pole concept is twice as much as the switching frequency, whereas in a non-pole commutated concept the sampling and switching frequencies are equal. This results in a better AC waveform quality and a smaller amplitude for higher order harmonics in commutated pole topologies. The non-pole commutated topologies need one or two semiconductor switches at their auxiliary circuit, but commutated pole topologies require six semiconductor switches at the auxiliary circuit. This is disadvantageous from the viewpoint of converter costs and circuitry except in high power applications. (see also section 2.4.)

Chapter 3: These generalized converter representations are used to drive design equations of the circuits and to explore the expressions giving the necessary boost energy for Half Wave-topologies. In a comparison among the switching and auxiliary circuit losses of AC-Side Soft Commutated (ASSC) PWM converters it is shown that the AC-Side Half-Wave Auxiliary Commutated Pole (ACS-HW-ARCP) converter (the slightly modified ARCP converter) has the lowest total losses due to ZCS conditions at its auxiliary circuit. (see also section 3.4.)

Chapter 4: A generalized space vector PWM (SVPWM) technique is introduced as Minimum-Loss Space Vector PWM (ML-SVPWM) for commutated pole converters. This technique minimizes the switching and auxiliary circuit losses of converter for a relatively wide range of variation of current phase angle φ. Compared to conventional continuous modulation under the same switching frequency, the results of loss calculations reveal that using the Minimum-Loss Space Vector PWM (ML-SVPWM) technique a reduction of 60% in both switching and auxiliary circuit losses can be achieved for the range $\varphi \in [-\frac{\pi}{6}, \frac{\pi}{6}]$. For the ranges $\varphi \in [-\frac{\pi}{3}, -\frac{\pi}{6}]$ and $\varphi \in [\frac{\pi}{6}, \frac{\pi}{3}]$ this reduction depends on φ. The results of calculation of the RMS value of current ripples of different SVPWM techniques show that under the same switching losses (different switching frequency) the Minimum-Loss Space Vector PWM (ML-SVPWM) technique is superior to other SVPWM techniques from the viewpoint of quality of AC current waveform for the entire range of modulation index. (see also section 4.6.)

Chapter 5: A detailed study of the influence of different stages of Zero Voltage Transition (ZVT) on the DC bus utilization of converters and on the linearity range of modulation is carried out in this work. The results show that duration of voltage transitions reduces the DC bus utilization. This becomes more considerable at high switching frequencies. A method, which can be easily implemented, is suggested for compensating this reduction. The effectiveness of the method is shown by simulated results. It is also shown that the duration of charging and discharging stages of ZVT limits the minimum duration of zero and non-zero voltage vectors leading to a reduction in the linearity range of modulation. For this reason the maximum linear modulation index of hard commutated converters, $m_{max} = 1.15$, cannot be achieved by soft commutated PWM converters. The expressions describing the dependence of linearity range of modulation on the auxiliary resonant network characteristics, modulation frequency and commutated current are derived for commutated pole topologies. It is shown that the Quarter Wave-concept and the discontinuous modulation techniques make a higher maximum linear modulation index available. (see also section 5.3.)

Chapter 6: Selecting a certain modulation technique for minimizing the losses, the current control techniques which can facilitate both the converter and selected modulation technique are strongly limited. The commonly used direct current control techniques cannot be used, because the converter switches should be now controlled according to the switching pattern of the selected modulation technique. One of the applicable current control methods could be the Predictive Dead-Beat Current Control Method which does not limit the switching pattern used. This method is analysed in detail in steady state and transient state. A DSP-based controller using this method is also described for AC-Side Soft Commutated Pole PWM Three-Phase converters in this chapter.

References

[1] Swanepoel, P.H. and Van Wyk, J.D.
Analysis and Optimization of Regenerative Linear Snubbers Applied to Switches with Voltage and Current Tails.
IEEE Transactions on Power Electronics, Vol. 9, No. 4, pp 433-442, July 1994.

[2] Undeland, T.; Jenset, F.; Steinbakk, A.; Rogne, T. and Hernes, M.
A Snubber Configuration for both Power Transistor and GTO PWM Inverters.
IEEE PESC Conference Records, pp 42-53, 1984.

[3] Blaabjerg, F.
Snubber in PWM-VSI Inverters.
IEEE PESC Conference Records, pp 104-111, 1991.

[4] Elasser, A. and Torrey, D.A.
Soft Switching Active Snubbers for DC/DC Converters.
IEEE Transactions on Power Electronics, Vol. 11, No. 5, pp 710-722, September 1996.

[5] Xiangning, H.; Finney, S.J. and Qian, Z.M.
Novel Passive Lossless Turn-on Snubber for Voltage Source Inverters.
IEEE Transactions on Power Electronics, Vol. 12, No. 1, pp 173-179, January 1997.

[6] McMurry, W.
Effivient Snubbers for Voltage-Source GTO Inverters.
IEEE Transactions on Power Electronics, Vol. 2, No. 3, pp 264-272, July 1987.

[7] Holtz, J.; Salama, S. and Werner, K.H.
A Nondissipative Snubber Circuit for High Power GTO Inverters.
IEEE Transactions on Industrial Applications, Vol. 25, No. 4, pp 620-626, July/August 1989.

[8] Teigelkötter, J.
Schaltverhalten und Schutzbeschaltungen von Hochleistungshalbleitern.
Ph.D. Thesis (VDI No. 206), Düsseldorf 1996.

[9] Schröder D.
Elektrische Antriebe 1, Grundlagen.
Springer–Verlag, Berlin 1994.

[10] Schröder D.
Elektrische Antriebe 2, Regelung Elektrischer Antriebe.
Springer–Verlag, Berlin 1995.

[11] Schröder D.
Elektrische Antriebe 3, Leistungselektronische Bauelemente.
Springer–Verlag, Berlin 1995.

[12] Schröder D.
Elektrische Antriebe 4, Leistungselektronische Schaltungen.
Springer–Verlag, Berlin 1998.

[13] Ziogas, P.D.; Ranganathan, V.T. and Stefanovic, V.
A Four Quadrant Current Regulated Converter With High Frequency Link.
IEEE Transactions on Industrial Applications, Vol. 18, pp 499-506, July./Aug. 1979.

[14] Schwarz, F.C. and Benklassens, J.
A Controllable 45 kW Current Source for DC Mashines.
IEEE Transactions on Industrial Applications, Vol. 15, pp 437-444, September/October 1982.

[15] Chin, Y. and Lee, F.C.
Constant Frequency Paeallel-Resonant Converter.
IEEE IAS Annual Conference Records, pp 705-710, 1987.

[16] Ngo, K.D.T.
Analysis of a Series Resonant Converter Pulse Witdh Modulation or Current-Controlled for Low Switching Losses.
IEEE IAS Annual Conference Records, pp 527-536, 1987.

[17] Murai, L. and Lipo, T.A.
High Frequency Series Resonant DC Link Power Conversion.
IEEE IAS Annual Conference Records, pp 772-779, 1988.

[18] Munk-Nielson, S.; Blaabjerg, F. and Pederson, J.K.
Link Voltage Peak control of Parallel Resonant converter by Control of the Converter Switching Instant.
IEEE PESC Conference Records, pp 929-935, 1995.

[19] Keller C.
Einsatzkriterien schneller abschaltbarer Leistungshalbleiter in Quasiresonanz-Umrichtern.
Ph.D. Thesis (D 83), Berlin 1991.

[20] Divan, D.M.
Resonant DC Link Converter - a New Concept in Static Power Conversation.
IEEE Industry Application Society Conference Proceeding, pp 267-280, 1986.

[21] Divan, D.M. and Skibinski, G.
Zero-Switching Loss Inverters for High-Power Applications.
IEEE Transactions on Industry Applications, Vol. 25, No. 4, pp 634-643, July/August 1989.

[22] Malesani, L.; Tenti, P.; Divan, D.M. and Toigo, V.
A Synchronized Resonant DC Link Converter for Soft Switched PWM.
IEEE Industry Application Society Conference Proceeding, pp 1037-1044, 1990.

[23] Sommer R.
Der Active Clamped Resonant DC-Link Inverter - eine resonante Umrichterschaltung zur Frequenzumformung.
Ph.D. Thesis (D 83), Berlin 1994.

[24] Chen, S. and Lipo, T.
A Novel Soft-Switched PWM Inverter for AC Motor Drives.
IEEE Transactions on Power Electronics, Vol. 11, No. 4, pp 653-659, July 1996.

[25] Choi, J. and Sul, S.
Resonant Link Bidirectional Power Converter: Part I-Resonant Circuit.
IEEE Transactions on Power Electronics, Vol. 10, No. 4, pp 479-484, July 1995.

[26] He, J. and Mohan, N.
Parallel Resonant DC Link Circuit-A Novel Zero Switching Loss Technology with Minimum Voltage Stresses.
IEEE PESC Conference Records, pp 1006-1012, 1989.

[27] Jung, Y.C.; Liu, H.L.; Cho, G.C. and Cho, G.H.
Soft Switching Space Vector PWM Inverter Using a New Quasi-Parallel Resonant DC Link.
IEEE Transactions on Power Electronics, Vol. 11, No. 3, pp 503-511, May 1996.

[28] Malesani, L.; Tenti, P. and Toigo, V.
High Efficiency Quasi Resonant DC Link Converter for Full-Range PWM.
IEEE APEC Conference Records, pp 472-478, 1992.

[29] Wang, K.; Jiang, Y.T.; Dubovsky, S.; Hua, G; Borojević, D. and Lee, F.C.
Novel DC Rail Soft Switched Three Phase Voltage Source Inverters.
IEEE Transactions on Industrial Applications, Vol. 33, No. 2, pp 509-517, March/April 1997.

[30] Yi, W.; Liu, H.L.; Jung, Y.C.; Cho, J.G. and Cho, G.C.
Program-controlled Soft Switching PRDCL Inverter with New Space Vector PWM algorithm.
International J. Electronics, Vol. 75, No. 3, pp 567-575, 1993.

[31] De Doncker, R.W. and Lyons, J.P.
The Auxiliary Resonant Commutated Pole Inverter.
IEEE Industry Applications Society Conference Records, pp 1228-1235, 1993.

[32] Gester, C. and Hofer, P.
Gate-controlled dv/dt- and di/dt-limitation in High Power IGBT Converters.
EPE Jurnal, Vol. 5, No. 3/4, pp 11-16, January 1996.

[33] Habetler, T.G.
A Space Vector-Based Rectifier Regulator for AC/DC/AC Converters.
IEEE Transactions on Power Electronics, Vol. 8, No. 1, pp 30-36, 1993.

[34] Hua, G.; Leu, C. and Lee, F.C.
Novel Zero-Voltage Transition PWM Converters.
IEEE PESC Conference Records, pp 1228-1235, 1992.

[35] Hua, G.; Yang, E.x.; Jiang, Y.; Leu, C. and Lee, F.C.
Novel Zero-Current Transition PWM Converters.
IEEE PESC Conference Records, pp 56-61, 1993.

[36] Cuadros, C.; Borojević, D.; Gatarić, S.; Vlatković, V. and Lee, F.C.
Space Vector Modulated, Zero-Voltage Transition Three-Phase to DC Bidirectional Converter.
VPEC Publication Series, Vol. V, pp 233-240, 1994.

[37] Vlatković, V.; Borojević, D. and Lee, F.C.
A New Zero-Voltage Transition, Three-Phase PWM Rectifier/Inverter.
IEEE PESC Conference Records, pp 868-873, 1993.

[38] Analogy Co.
Saber Simulation Reference Manual for Release 4.0.
SSR 4-105, SSR 4-111, SSR 4-112; AGD 8-2, , 1995.

[39] Vogler, T. and Schröder, D.
A New and Accurate Circuit Modeling Approach for the Power Diode.
IEEE PESC Conference Records, pp 871-876, 1992.

[40] Metzner, D.; Vogler, T. and Schröder, D.
A Modulator Concept for the Circuit Simulation of Bipolar Power Semiconductors.
EPE Conference Records, Vol. 3, pp 15-22, 1993.

[41] Metzner, D.
Netzwerkmodelle abschaltbarer Leistungshalbleiter-Bauelemente.
Ph.D. Thesis, TUM Munich 1994.

[42] Hefner, Jr., A.R. and Daniel, M.D.
An Experimentally Verified IGBT Model Implemented in the Saber Circuit Simulator.

IEEE Transactions on Power Electronics, Vol. 9, No. 5, pp 532-542, September 1994.

[43] Hefner, Jr., A.R. and Daniel, M.D.
Modeling Buffer Layer IGBT's for Circuit Simulation.
IEEE Transactions on Power Electronics, Vol. 10, No. 2, pp 111-123, March 1995.

[44] De Doncker, R.W.
Soft-Switching Turn-off Behavior Models.
EPE Association Tutroial: Resonant Converters, pp 265-270, Sep. 1993.

[45] Heck A.
Introduction to Maple
Springer-Verlag 1993.

[46] Höringer M.
Maple in Technik und Wissenschaft
Addison-Wesley 1996.

[47] International Rectifier
IRGPH50KD2 Data Sheet
The key to New Designs, IGBTs, pp 30-36, 1993.

[48] Siemens
BSM 191(C) Data Sheet
SIPMOS Halbleiter Data Book, pp 227, 1991/92.

[49] International Rectifier
12FL60S02 Data Sheet
Power Semiconductor,Product digest, pp M-3, 1994.

[50] International Rectifier
IRGPH40F Data Sheet
IGBT Designer's Manual, pp C,273-C,278, 1994.

[51] International Rectifier
IRGBC30UD2 Data Sheet
IGBT Designer's Manual, pp C,701-C,708, 1994.

[52] Dehmlow, M.
Vergleichende Betrachtung unterschiedlicher resonanter Umrichtertopologien.
Ph.D. Thesis (D 83), Berlin 1995.

[53] Protiwa, F.F.
Vergleich dreiphasiger Resonanzwechselrichter in Simulation und Messung.
Ph.D. Thesis , Aachen 1996.

[54] Radan, A. and Schröder, D.
Study of Three-Phase AC-Side Soft Commutated Resonant Pole PWM Conver-

ters.
PCIM Conference Records, Submitted Paper, June 2000.

[55] Kaura, V. and Blasko, V.
Operation of a Voltage Source Converter at Increased Utility Voltage.
IEEE Transactions on Power Electronics, Vol. 12, No. 1, pp 132-137, January 1997.

[56] Bowes, S.R.
New Sinusiodal Pulse Width Modulated Inverter.
Proceeding IEE, Vol. 122, No. 11, pp 1279-1285, 1975.

[57] Bowes, S.R. and Clemens R.R.
Computer Aided Design of PWM Inverter Systems.
Proceeding IEE B, Elect. Power Appl., Vol. 129, No. 1, pp 1-17, 1982.

[58] Boost, M.R.
State-Of-The-Art Carrier PWM Techniques: A Critical Evaluation.
IEEE Transactions on Industry Applications, Vol. 24, pp 271-280, March/April 1988.

[59] Qing, Z.; Zhihui, L. and Chen, J.
A Novel Three Phase SPWM Technique: Instantaneously Floating The Neutral Point.
PEDS Conference Records, pp 781-786, 1997.

[60] Pfaff, G. and Wick, A.
Direkte Stromregelung bei Drehstromantrieben mit Pulswechselrichter.
Regelungstechnische Praxis 24, Heft 11, Jahresgang 1983.

[61] Van Der Broeck, H.W. and Skudelny, H.C.
Analysis and Realization of a Pulse Width Modulator Based on Voltage Space Vectors.
IEEE Transactions on Industry Applications, Vol. 24, No. 1, pp 142-149, January/February 1988.

[62] Busse, H. and Holtz, J.
Novel Multiloop Control of a Unity Power Factor Fast Switching AC to DC Converter.
IEEE PESC Conference Records, pp 171-179, 1982.

[63] Jenni, F. and Wüst, D.
Steuerverfahren für selbstgeführte Stromrichter.
vdf Hochschulverlag AG an der ETH Zürich - B.G. Taubner Stuttgart, pp 152-155, 1995.

[64] Fukuda, S. and Iwaji, W.
A Pulse Frequency Modulated PWM Scheme for Sinusoidal inverters.
IEEE PESC Conference Records, pp 382-389, 1991.

[65] Trzynadlowski, A.M.; Kirlin R.L. and Legowski, S.F.
Space Vector PWM Technique with Minimum Switching Losses and Variable Pulse Rate.
IEEE Transactions on Industrial Electronics, Vol. 44, No. 2, pp 173-181, January 1997.

[66] Sun, J. and Grotstollen, H.
Optimized Space Vector Modulation and Regular-Sampled PWM: A Reexamination.
IEEE Industry Application Society Conference Proceeding, pp 956-963, 1996.

[67] Blasko, V.
A Hybrid PWM Strategy Combining Modified Space Vector and Triangle Comparison Methods.
IEEE PESC Conference Records, pp 1872-1878, 1996.

[68] Blasko, V.
Analysis of a Hybrid PWM Based Modified Space Vector and Triangle Comparison Methods.
IEEE Transactions on Industry Applications, Vol. 33, No. 3, pp 756-764, May/June 1997.

[69] Bolognani, S. and Zigliotto M.
Novel Digital Continuous Control of SVM Inverters in the Overmodulation Range.
IEEE Transactions on Industry Applications, Vol. 33, No. 2, pp 525-530, March/April 1997.

[70] Depenbrock, M.
Pulse Width Control of a Three Phase Inverter With Nonsinusoidal Phase Voltage.
IEEE ISPC Conference Records, pp 399-403, 1977.

[71] Kim, J.S. and Sul S.K.
A Novel Voltage Modulation Technique of the Space Vector PWM.
IPEC Conference Records, pp 742-747, 1995.

[72] Ogasawara, S.; Akagi, H. and Nabae, A.
A Novel PWM Scheme of Voltage Source Inverters Based on Space Vector Theory.
EPE Conference Records, pp 1197-1202, 1989.

[73] Lai, Y.S. and Bowes, S.R.
A Universal Space Vector Modulation Strategy Based on Regular-Sampled Pulse Width Modulation.
IEEE APEC Conference Records, pp 120-126, 1996.

[74] Hava, A.M. Kerkman, R.J. and Lipo, T.A.
A High Performance Generalized Discontinuous PWM Algorithm.
IEEE APEC Conference Records, pp 886-894, 1997.

[75] Houldsworth, J.A. and Grant, D.A.
The Use of Harmonic Distortion to Increase the Output Voltage of a Three Phase PWM Inverter.
IEEE Transactions on Industry Applications, Vol. 20, No. 5, pp 1224-1228, September/October 1984.

[76] Van Der Bröck, H.
Analysis of the Harmonics in Voltage Fed Inverter Drives Caused by PWM Schemes with Discontinuous Switching Operation.
EPE Conference Records, Vol. 3, pp 261-266, 1991.

[77] Perruchoud, P.J.P. and Pinewski, P.
Power Losses for Space Vector Modulation Techniques.
IEEE Power Electronics in Transportation Cenference Proceeding , pp 167-173, 1996.

[78] Alexander, D.R. and Williams, S.P.
An Optimal PWM Algorithm Implementation in a High Performance 125 KVA Inverter.
—— , pp 771-777

[79] Malesani, L.; Tomasin, P. and Toigo, V.
Space Vector Control and Current Harmonics in Quasi-Resonant Soft-Switching PWM Converssion.
IEEE Transactions on Industry Applications, Vol. 32, No. 2, pp 269-277, March/April 1996.

[80] Chung, D. and Sul, S.
Minimum-Loss PWM Strategy for 3-Phase Rectifier.
IEEE PESC Conference Records, 1997.

[81] Kolar, j.W.; Ertl H. and Zach F.
Influence of the Modulation Method on the Conduction and Switching Losses of a PWM Converter System.
IEEE Transactions on Industry Applications, Vol. 27, No. 6, pp 1063-1075, November/December 1991.

[82] Kolar, j.W.; Ertl H. and Zach F.
Minimizing the Current Harmonics RMS Value of Three-Phase PWM Converter System by Optimal and Suboptimal Transition Between Continuous and Discontinuous Modulation.
IEEE PESC Conference Records, pp 372-381, 1991.

[83] Mazumder, S.
Complete Mathematical Analysis of Current Ripple as a Function of Modulation Index for Direct Indirect and Bus Clamped Space Vector Modulation Techniques.
IEEE Int. Elec. Machines and Drives Conference Records, pp TC-10.1-3, 1997.

[84] Holtz, J. and Stadtfeld, S.
A Predictive Controller for the Stator Current Vector of AC Machines Fed From a Switched Voltage Source.
IEEE International Power Electronics Conference, Record, pp 1665-1675, 1983.

[85] Pfaff, G.; Weschta A. and Wick A.
Design and Experiment Results of a Brushless AC Servo-Drive.
IEEE Industry Applications Society Annual Meeting Record, pp 692-697, 1992.

[86] Nabae, A.; Ogasawara S. and Akagi H.
A Novel Control Scheme of Current Controlled PWM Inverters.
IEEE Industry Applications Society Annual Meeting Record, pp 473-478, 1985.

[87] Wüst D. and Jenni F.
Space Vector Based Current Control Schemes for Voltage Source Inverters.
IEEE PESC Conference Records, pp 986-992, 1993.

[88] Kazmierkowski, M.P.; Dzieniakowski, M.A. and Sulkowski, W.
Novel Space Vector Based Current Controller for PWM Inverters.
IEEE Transactions on Power Electronics, Vol. 6, No. 1, pp 158-165, 1991.

[89] Kazmierkowski, M.P. and Dzieniakowski, M.A.
Review of Current Regulation Techniques for Three-Phase PWM Inverters.
IEEE Conference on Industrial Electronics, Control and Instrumentation, Record, pp 567-575, 1994.

[90] Kolar, j.W.; Ertl H. and Zach F.
Analysis of On- and Off-Line Optimized Predictive Current Controllers for PWM Converter System.
IEEE Transactions on Power Electronics, Vol. 6, No. 3, pp 451-462, 1991.

[91] Mohan, N.; Undeland, T.M. and Robbins, W.P.
Power Electronics: Converters, Applications and Design.
John Wiley & Sons 1989.

A. Soft Switching Turn-off Behavior Models

Four different device turn-off behavior models are introduced and analysed in [44] which are useful for comparative analysis and simulations. In this section two models used in this work are described.

For the analysis of circuits, the components are assumed to be lossless passive components, the current I_{off} is assumed to be constant during the device turn-off transition and the snubber capacitance voltage is assumed not to reach the DC bus voltage within the turn-off time interval of the device (see Fig. A.1).

Under the assumption given above the soft switching turn-off energy of a device can be calculated from the simple circuit shown in Fig. A.1. The two models of switch turn-off current used in this work are shown in Fig. A.2.

In the linear fall model, the current in the devices decays linearly and reaches zero during the time t_f. The current in the capacitor I_{C_S} is complementary to the device current I_{AK} and increases linearly according to:

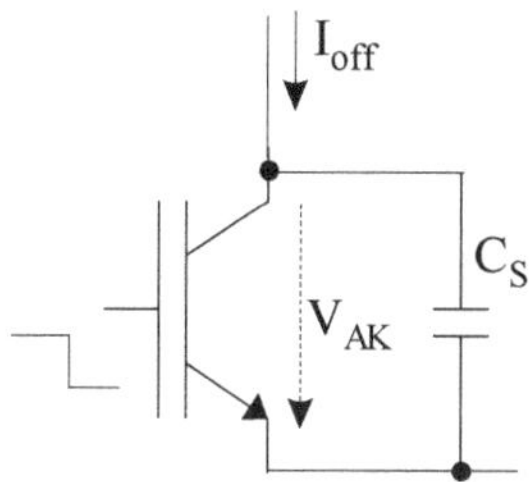

Fig. A.1: Soft switching turn-off IGBT

$$I_{C_S}(t) = I_{off}\,\frac{t}{t_f} \qquad 0 \le t \le t_f. \tag{A.1}$$

The voltage across the device V_{AK} increases according to:

$$V_{AK} = \frac{1}{C_S}\int_0^t I_{C_S}(t)dt = \frac{I_{off}}{2C_S}\frac{t^2}{t_f}. \tag{A.2}$$

Multiplying the device current I_{AK} with the voltage V_{AK} results in the instantaneous power loss in the device. Integrating this loss over the turn-off interval yields the turn-off energy given by:

$$E_{off} = \frac{I_{off}^2\,t_f^2}{24C_S}. \tag{A.3}$$

This model is more applicable when devices without tail current such as Power MOSFETs are applied. This model is used in this work to calculated the turn-off

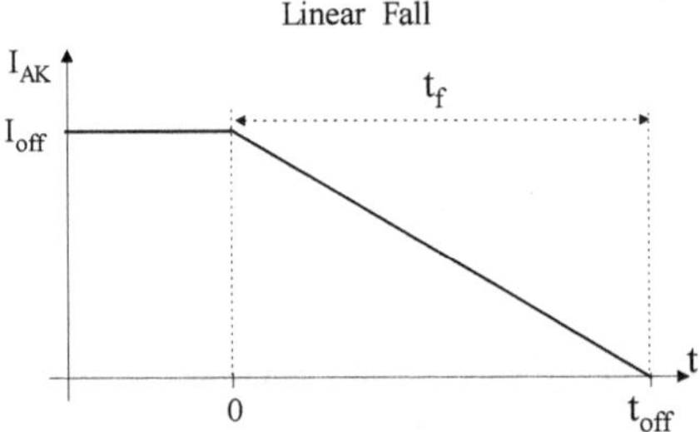

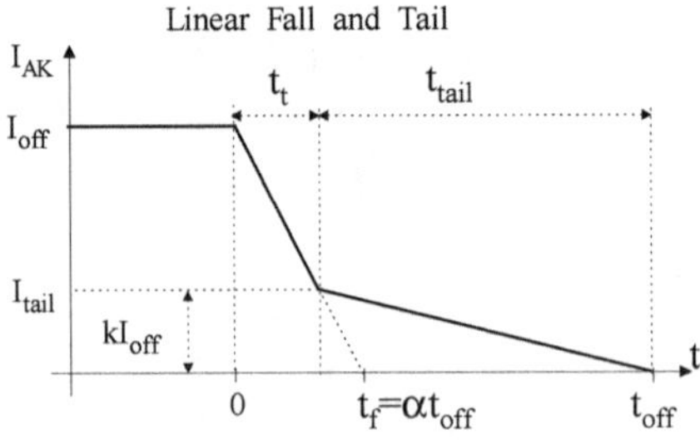

Fig. A.2: Turn-off behavior models

loss of Power MOSFETs.

The second model comprises a linear tail model and is called linear fall and tail model. Since this model offers an adequate accuracy and can be fitted closely to the usual behavior of IGBTs, it is used in this work for calculation of the turn-off losses of these power devices. The expression of turn-off energy is calculated in the same way as described above and is given as follows:

$$\begin{aligned}
E_{off} &= \frac{I_{off}^2 t_{off}^2}{24C_S}\zeta \qquad (A.4)\\
\zeta &= (1-a)^2 k(1-k) + a^2(1-k)(1+3k) + 6a(1-a)k(1-k)\\
t_{off} &= t_t + t_{tail}\\
a &= \frac{t_t}{t_{off}}\\
k &= \frac{I_{tail}}{I_{off}}
\end{aligned}$$

B. Normalization and Fourier Analysis of AC Currents

For normalizing the AC voltages, $V_B = \frac{V_{dc}}{2}$, is selected. According to the definition of modulation index, m:

$$m = \frac{|V_{ref}|}{\frac{V_{dc}}{2}}, \tag{B.1}$$

V_B corresponds to the amplitude of fundamental component (ω_1) of inverter phase voltage when $m = 1$. With normalizing the AC voltages to this value, the amplitude of the normalized reference phase voltage $\hat{v}_{ref}$, which corresponds to the amplitude of the fundamental component of the normalized inverter phase voltage $\hat{v}_{a(1)}$ (see Fig. B.1), also represents the modulation index m. The base power for the results shown in chapter 4 is considered to be $P_B = V_B\, I_B = V_B^2$ resulting in $Z_B = 1$. Hence, the base current $I_B = V_B$ also represents the amplitude of the fundamental component (ω_1) of inverter phase current at $m = 1$

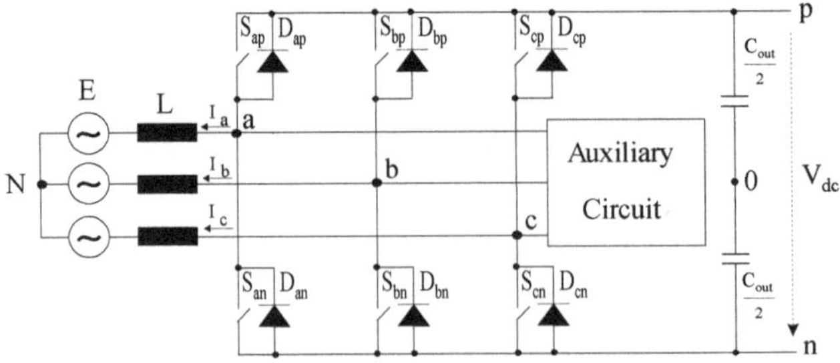

Fig. B.1: Three-Phase Inverter Circuit

and the amplitude of the normalized inverter phase current $\hat{i}_{a(1)}$ corresponds to the modulation index m also.

The inverter phase current is calculated by:

$$I_a(t) = \frac{1}{L} \int \Big[V_a(t) - E_a(t)\Big]\, dt. \tag{B.2}$$

Normalizing this equation results in:

$$i_a(t) = \omega_1 \int \Big[v_a(t) - e_a(t)\Big]\, dt \tag{B.3}$$

which is independent from L. The normalized phase current is calculated by the SABER simulation program using above equation. Using Fourier Analyser of this simulator, the frequency spectrum of the normalized phase current is calculated for the different modulation techniques. The Simulator also calculates the Total Harmonic Distortion (THD) of AC waveforms using following equation [38]:

$$THD = \frac{\sqrt{\sum\limits_{i=2}^{\infty} i_{a(i)}}}{\sqrt{\sum\limits_{i=1}^{\infty} i_{a(i)}}}. \tag{B.4}$$

The duty cycle of all voltage vectors is also normalized to the sampling cycle T_S to represents their duty ratio.

C. Symbols

Electrical values

γ	apportioning factor of duration of zero voltage vectors
Δt	necessary duration for holding the ZVS conditions across the main switches
$\Delta t_{con,s}$	conduction duration of auxiliary switch during a zero voltage transition
$\Delta t_{con,d}$	conduction duration of auxiliary diode during a zero voltage transition
Δt_{ch}	duration of charging state of auxiliary circuit
Δt_{res}	duration of resonant state of auxiliary circuit
Δt_{dis}	duration of discharging state of auxiliary circuit
Δt_{ZVS}	total duration of zero voltage transition
ΔI_K	current deviation caused by switching the voltage vector $\vec{V}_K$
ΔI^2_{rms}	square RMS value of current ripples
$\vec{A}_{VS(ref)}$	desired reference volt-second area vector
$\vec{A}_{VS(res)}$	volt-second area vector during the resonant state
$\vec{A}_{VS(sc)}$	volt-second area vector during the current self-commutation
C_{out}	converter DC-side capacitance
E	voltage source at the AC-side of converter
$E_{con,IGBT}$	conduction energy loss of IGBTs during $\Delta t_{con,s}$
$E_{con,d}$	conduction energy loss of auxiliary diode during $\Delta t_{con,d}$
$E_{con,S_{ap}}$	conduction energy loss of the switch S_{ap}
$E_{con(aux,S)}$	conduction energy loss of the auxiliary switches
$E_{con(aux,D)}$	conduction energy loss of the auxiliary diodes
$E_{off,IGBT}$	turn-off energy loss of IGBT

$E_{off,MOS}$	turn-off energy loss of Power MOSFET
$E_{on,MOS}$	capacitive turn-on energy loss of Power MOSFET
$\mathcal{E}_{off}$	normalized per phase turn-off energy loss indicator
f_1	fundamental frequency
f_S	sampling frequency
f_{sw}	switching frequency
I_{boost}	boost current
K	number of the voltage vector in which $\vec{V}_{ref}$ is located
m	modulation index
$P_{ref(out)}$	power transferred by converter
R_X	ratio of the global per phase turn-off energy loss caused by the modulation technique X to that caused by the C-SVPWM technique
T_1	fundamental cycle and duty-cycle of voltage vector $\vec{V}_1$
t_a	normalized duty cycle of phase a
T_K	duty cycle of voltage vector $\vec{V}_K$
t_K	normalized duty cycle of voltage vector $\vec{V}_K$
T_S	sampling cycle and base value for normalization of duty cycle of voltage vectors
T_r	resonance cycle
T_{sw}	switching cycle
V_{dc}	converter DC link voltage
$V_{F,dev}$	forward voltage drop across devices
$\vec{V}_{res,abc}$	node voltage vector during the resonant state in three-phase coordination system
$\vec{V}_{sc,abc}$	node voltage vector during the current self-commutation in three-phase coordination system
$\vec{V}_{res,\alpha\beta}$	node voltage vector during the resonant state in stationary $\alpha\beta$ reference frame
$\vec{V}_{sc,\alpha\beta}$	node voltage vector during the current self-commutation in stationary $\alpha\beta$ reference frame
V, I	Instantaneous value of variable
$\hat{V}$, $\hat{I}$	Amplitude of sinusoidal variable
v, i	Normalized variable
$\bar{v}$, $\bar{i}$	locally (with respect to T_S) averaged normalized variable
V_{N0}	voltage between the AC-side neutral point and the center point of converter DC-side capacitance

v_{zs}	normalized zero sequence voltage
ω_1	fundamental angular frequency
ω_r	angular frequency of auxiliary resonant circuit
Z_r	impedance of auxiliary resonant circuit

Angles

ϕ	phase angle of $\vec{V}_{ref}$
φ	phase difference angle between the phase voltage and current
ψ	modulation angle (shifting phase angle of zero sequence voltage)
θ	switching sequence changing angle
α_d	the angle within which only discontinuous modulation technique can be used
α_c	the angle within which only continuous modulation technique can be used
ε_0	zero voltage vector elimination angle

Indices

$\_^S$	quantity in the stationary $\alpha\beta$ reference frame
$\_^K$	quantity in the dq reference frame rotating with the frequency ω_1 and aligned with E_a
$-_1$	fundamental quantity
$-_B$	Base value
$-_{ref}$	reference quantity
$-_{max}$	absolute maximum value (part) of a quantity (waveform)
$-_{min}$	absolute minimum value (part) of a quantity (waveform)
$-_{mid}$	middle part of a waveform (see Fig. 4.8(a))
$-_{ph}$	phase quantity
$-_X$	quantity related to modulation technique X
$-_{HW}$	quantity related to HW concept
$-_{QW}$	quantity related to QW concept
$-_{a,b,c}$	quantity related to phase a, b or c in three phase system
$-_{\alpha,\beta}$	quantity related to α- or β-component in stationary $\alpha\beta$ reference frame
$-_{d,q}$	quantity related to d- or q-component in rotating dq reference frame
$-_{xa,xb,xc}$	quantity related to the auxiliary branch at phase a, b or c

$-_{cm}$	quantity related to continuous modulation
$-_{dm}$	quantity related to discontinuous modulation
$-_{local}$	local (with respect to T_S) mean/RMS value of variable
$-_{global}$	global (with respect to T_1) mean/RMS value of variable

D. Glossary

Zero Switching States: From the eight possible switching combinations of main switches of a three-phase converter, $S_{0(nnn)}$, $S_{1(pnn)}, \cdots, S_{7(ppp)}$, the switching states $S_{0(nnn)}$ and $S_{7(ppp)}$ generate zero voltage at the AC-side of converter and are called Zero Switching States. In the switching state $S_{0(nnn)}$ the three bottom switches S_{an}, S_{bn} and S_{cn} are closed and all the three phases are clamped to the negative DC rail n (see Fig. D.3). In the switching state $S_{7(ppp)}$ the three upper switches S_{ap}, S_{bp} and S_{cp} are closed and all the three phases are clamped to the positive DC rail p (see also Fig. 2.3).

Non-Zero Switching States: The switching states $S_{1(pnn)}$, $S_{2(ppn)}, \cdots, S_{6(pnp)}$ are called Non-Zero Switching States because they generate non-zero voltage at the AC-side of converter (see Fig. 2.3).

Zero Voltage Vector: The voltage vectors $\vec{V}_0$ and $\vec{V}_7$ generated by zero switching states $S_{0(nnn)}$ and $S_{7(ppp)}$ are called Zero Voltage Vectors.

Non-Zero Voltage Vector: The voltage vectors $\vec{V}_1$, $\vec{V}_2, \cdots, \vec{V}_6$ generated by non-zero switching states $S_{1(pnn)}$, $S_{2(ppn)}, \cdots, S_{6(pnp)}$ are called Non-Zero Voltage Vectors.

HW and QW Concepts: The topologies whose resonant network creates the zero voltage switching conditions for soft switching of main switches after a Half-Wave (HW) resonating are named HW-Concept. QW-concept, in contrary, creates these conditions after a Quarter-Wave (QW) resonating (see Fig. D.1).

Self-Commutation: Soft switch-to-diode Current Commutation is called a Self-Commutation because the soft transfer of phase current from one switch to the diode located at the opposite side of the same leg of converter occurs naturally without an auxiliary circuit action in case the phase

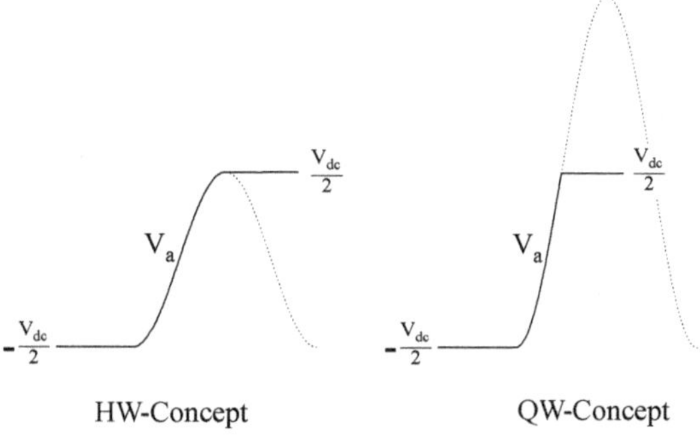

Fig. D.1: HW- and QW-concept.

current is large enough. For example, suppose the upper switch, S_{ap} in Fig. D.2 (left), at phase a conducting phase current I_a and I_a is flowing in the direction shown. If switch S_{ap} is turned off, I_a automatically starts to charge the capacitance across this switch and to discharge the capacitance across diode D_{an}. The node voltage V_a softly decreases. When V_a decays to zero, D_{an} starts conducting the phase current I_a. So the commutation of I_a from switch S_{ap} to diode D_{an} occurs naturally and has no need for an auxiliary circuit action. Because of similarity of this kind of current commutation with that of line-frequency phase-controlled thyristor rectifiers [91], it is called self-commutation in this work.

ZVS-Commutation: Soft diode-to-switch current commutation and Soft switch-to-diode Current Commutation in case of small phase currents are called ZVS-commutations in this work. This kind of current commutations always need the auxiliary circuit action to soften the conditions of commutation. An example of ZVS-commutation is shown in Fig. D.2(right). Without activating the auxiliary circuit the commutation will be a hard one and causes problems related to hard switching described in chapters 1 and 2.

Phase Duty Cycle: Total duration in which the respective phase is clamped to the positive DC rail within one sampling cycle is named phase duty cycle.

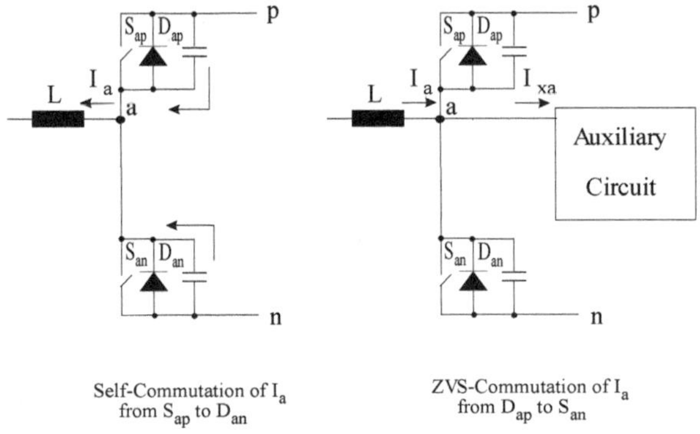

Fig. D.2: Different kinds of current commutation.

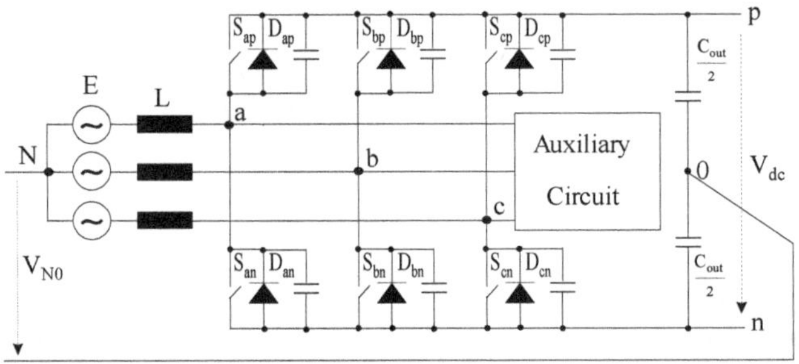

Fig. D.3: Three-Phase Inverter Circuit.

Common Mode Voltage: The star (neutral) point voltage of star connected loads at the AC side of three-phase converter referred to the center point of DC side, V_{N0}, in Fig. D.3 is named common mode voltage.

Zero Sequence Signal (Voltage): In this work the zero sequence signal (voltage) is referred to a signal (voltage) with the frequency $3 \times f_1$. In power system engineering the voltages with the frequency $(3K-2) \times f_1$, $(3K-1) \times f_1$ and $3K \times f_1$ $(K = 1, 2, 3 \cdots)$ are known as Positive Sequence Voltages, Negative Sequence Voltages and Zero Sequence Voltages. (see also Fig. D.4)

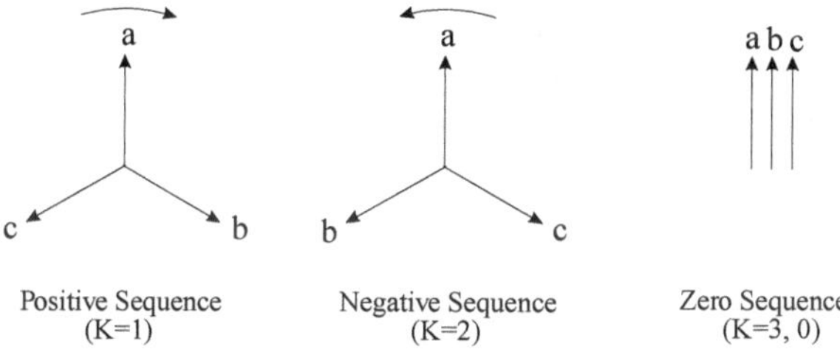

Fig. D.4: Three-Phase System.

E. List of Abbreviations

ACA: Auxiliary Circuit Action.

ASSC PWM: AC-Side Soft Commutated Pulse Width Modulated (converter).

ACS-HW-ARC: AC-Side Half-Wave Auxiliary Resonant Commutated (converter).

ACS-HW-ARCP: AC-Side Half-Wave Auxiliary Resonant Commutated Pole (converter).

ACS-QW-ARC: AC-Side Quarter-Wave Auxiliary Resonant Commutated (converter).

ACS-QW-ARCP: AC-Side Quarter-Wave Auxiliary Resonant Commutated Pole (converter).

C-SVPWM: Continuous Space Vector Pulse Width Modulation.

D-SVPWM: Discontinuous Space Vector Pulse Width Modulation.

D7-SVPWM: Discontinuous Space Vector Pulse Width Modulation (selecting only $\vec{V}_7$ as zero voltage vector).

D0-SVPWM: Discontinuous Space Vector Pulse Width Modulation (selecting only $\vec{V}_0$ as zero voltage vector).

D-SVPWM4: Discontinuous Space Vector Pulse Width Modulation (having four 30° no switching intervals distributed symmetrically over the fundamental cycle).

D-SVPWM5: Discontinuous Space Vector Pulse Width Modulation (having four 30° no switching intervals with changing the zero voltage vectors in every 30°).

HI-SPWM: Harmonic-Injected Sinusoidal Pulse Width Modulation.

ML-SVPWM: Minimum-Loss Space Vector Pulse Width Modulation.

SPWM: Sinusoidal Pulse Width Modulation.

SVPWM: Space Vector Pulse Width Modulation.

ST-SVPWM: Synchronous Turn-on Space Vector Pulse Width Modulation.

THD: Total Harmonic Distortion.

ZCS: Zero Current Switching.

ZVS: Zero Voltage Switching.

Zeitfracht Medien GmbH
Ferdinand-Jühlke-Straße 7
99095 Erfurt, Deutschland
produktsicherheit@kolibri360.de